Inter-Spacecraft Frequency Distribution

for Future Gravitational Wave Observatories

Von der QUEST-Leibniz-Forschungsschule der
Gottfried Wilhelm Leibniz Universität Hannover
zur Erlangung des Grades

Doktor der Naturwissenschaften
Dr. rer. nat.

genehmigte Dissertation

von
Dipl.-Phys. Simon Barke
geboren am 22. Oktober 1980
in Hannover, Deutschland

2015

Referent
Prof. Dr. Karsten Danzmann
Albert Einstein Institute (AEI)
Hanover, Germany

Korreferent
Prof. Dr. Guido Müller
University of Florida (UF)
Gainesville, FL, US

Korreferent
Prof. Dr. Claus Braxmaier
Center of Applied Space Technology and Microgravity (ZARM)
Bremen, Germany

Tag der Promotion
27. März 2015

DESIGN, DEVELOPMENT, AND TESTING OF A FREQUENCY GENERATION AND DISTRIBUTION SYSTEM FOR SPACEBORNE LASER INTERFEROMETRIC GRAVITATIONAL WAVE DETECTORS

2008 – 2014

Albert Einstein Institute (AEI), Hanover, Germany
Max Planck Institute for Gravitational Physics and
Leibniz Universität Hannover, Institute for Gravitational Physics

Centre for Quantum Engineering and Space-Time Research (QUEST),
Hanover, Germany

University of Florida (UF), Gainesville, Florida, USA
Department of Physics

in collaboration with

National Space Institute, Lyngby, Denmark
Technical University of Denmark

Axcon ApS, Lyngby, Denmark
The FPGA Power House

European Space Research and Technology Centre (ESTEC)
European Space Agency (ESA), Noordwijk, The Netherlands

Dedicated to Galileo Galilei,
who was sentenced to house arrest for the rest of his life by the
Catholic Church for publishing evidence and holding as true that the
Earth revolves around the Sun.

1564 − 1642

*By the grace of God [...] you, Galileo, son of the late Vincenzio
Galilei, Florentine, aged seventy years, were denounced to this
Holy Office in 1615 for holding as true the false doctrine taught
by some that the sun is the center of the world and motionless
and the earth moves even with diurnal motion.*

— The seven Cardinals of the Inquisition's Judgment in 1633 [1]

ABSTRACT

Gravitational waves hold the answer to some of the most fundamental questions about the universe and allow us to study as yet unobservable objects. Of the many detection principles, a laser interferometric gravitational wave observatory in space offers the most rewarding bandwidth with an abundance of astrophysical sources. Currently, the overall goal is to find the optimal concept for a gravitational wave observatory to be launched by the European Space Agency in the 2030s. This observatory needs to provide the best possible astrophysical value within given financial and technical constraints.

This thesis explains the influences of different mission parameters on the observatory's detection limit and presents a web application to quickly identify limiting noise sources. One aspect that is rarely thoroughly addressed by other studies is the required frequency range of the inter-spacecraft heterodyne signal. For the first time, this important subject is evaluated in detail for a wide range of interferometer arm lengths and mission lifetimes. Laser relative intensity noise imposes constraints on the lowest viable heterodyne frequency. The necessary upper end of the heterodyne frequency range was found to be between 10 and 28 MHz using a specially developed optimization algorithm. This newly acquired knowledge allows the improvement of parameter sets for currently considered mission concepts. Observatories seem to be possible that are less demanding on the instrument due to relaxed phase noise and timing stability requirements but still offer the same sensitivity to gravitational waves and may be even cheaper due to more compact telescopes. However, independent of the exact final mission design, the timing jitter of individual on-board reference oscillators needs to be below $\approx 4 \times 10^{-14}\,\mathrm{s}/\sqrt{\mathrm{Hz}}$ for frequencies down to almost $10^{-3}\,\mathrm{Hz}$. Such a stability is necessary to keep the metrology system noise significantly below the signal read-out noise of the received beam. Oscillators that stable do not exist.

For this reason, a system was designed that synchronizes all measurements performed on board the different spacecraft and thus makes the local ultrastable oscillators obsolete. This Inter-Spacecraft Frequency Distribution System generates a reference signal, converts it to different frequencies, and distributes it locally and throughout the entire constellation. Active and passive electronics such as frequency dividers and power splitters are involved, as well as electro-optics and optical components such as modulators and fiber amplifiers. For this system to work properly, every single component has to obey the demanding timing stability requirement, which was challenging to meet even for electrical cables. Finally, after five years of research, a TRL 4 compliant and fully operational Frequency Distribution System was constructed and successfully tested. Together with the overall metrology system, this item represented the only important technology relevant for gravitational wave observatories that was not available in Europe prior to this work. The developed prototype is applicable for a wide range of observatory concepts and will be refined within the next years to meet all demands of a future space mission.

Keywords: gravitational waves; laser interferometry; timing noise

KURZFASSUNG

Gravitationswellen enthalten die Antwort auf einige der grundlegendsten Fragen zu unserem Universum. Zudem erlauben sie es Objekte zu studieren, die bisher nicht zu beobachten sind. Unter den vielen Detektionsmethoden ermöglichen uns weltraumbasierte laserinterferometrische Gravitationswellenobservatorien Einblick in die ergiebigsten Frequenzbereich mit einer Vielzahl astrophysikalischer Quellen. Momentan besteht das generelle Ziel darin, das optimale Konzept für ein Gravitationswellenobservatorium zu finden, welches in den 2030er Jahren von der Europäischen Weltraumagentur gestartet werden soll. Unter Einbeziehung von finanziellen und technischen Einschränkungen muss dieses Observatorium eine bestmögliche astrophysikalische Ausbeute bieten.

Die vorliegende Arbeit erklärt die Einflüsse verschiedenster Missionsparameter auf die Detektionsgrenzen des Observatoriums und präsentiert eine Webapplikation, mit der limitierende Einflüsse schnell identifiziert werden können. Ein Aspekt, der nur selten ausgiebig in anderen Studien behandelt wird, ist die benötigte Frequenzbandbreite des interferometrischen Überlagerungssignals zwischen den verschiedenen Satelliten. Erstmals konnte dieses wichtige Thema im Detail für verschiedenste Armlängen und Missionsdauern untersucht werden. Die kleinstmögliche Heterodynfrequenz wird bestimmt von dem relativen Leistungsrauschen der Laser. Die nötige obere Grenze liegt je nach Missionskonzept zwischen 10 und 28 MHz und wurde über einen eigens entwickelten Optimierungsalgorithmus bestimmt. Das neu gewonnene Wissen erlaubt es, Missionsparameter für derzeit in Erwägung gezogene Missionskonzepte zu optimieren. Es scheint möglich, Observatorien zu entwerfen, die weniger strenge Phasenrausch- und Zeitstabilitätsanforderungen an die Instrumente stellen, aber gleichzeitig eine ebenso gute Gravitationswellenempfindlichkeit aufweisen und möglicherweise aufgrund kompakterer Teleskope sogar günstiger sind. Dennoch, unabhängig von dem endgültigen Missionsdesign benötigen die individuellen Satelliten Referenzoszillatoren, deren Zeitrauschen unterhalb von $\approx$ $4 \times 10^{-14}\,\mathrm{s}/\sqrt{\mathrm{Hz}}$ liegt – und dies hinab zu Frequenzen von fast 10^{-3} Hz. Eine solche Stabilität ist nötig, um das Rauschen des Metrologiesystems deutlich unter dem Ausleserauschen des eingehenden Signals zu halten. Derart stabile Oszillatoren existieren nicht.

Aus diesem Grund wurde ein System konzipiert, welches die Messungen an Bord der einzelnen Satelliten synchronisiert. So werden die lokalen hochstabilen Oszillatoren überflüssig. Dieses 'Inter-Spacecraft Frequency Distribution System' erzeugt ein Referenzsignal, konvertiert es in unterschiedliche Frequenzen und verteilt es sowohl lokal als auch zwischen den Satelliten. Das System besteht aus aktiven und passiven elektronischen, elektrooptischen und optischen Komponenten wie z.B. Frequenz- und Leistungsteiler, Modulatoren und Faserverstärker. Für eine einwandfreie Funktion muss jede einzelne Komponente die strengen Anforderungen an die Zeitstabilität erfüllen. Dabei stellten selbst elektrische Kabel eine Herausforderung dar. Am Ende konnte jedoch nach fünfjähriger Forschung ein TRL 4 konformes und voll betriebfähiges System zusammengestellt und erfolgreich getestet werden. Dieses System war zusammen mit dem allgemeinen Metrologiesystem die einzige wichtige Technologie relevant für Gravitationswellenobservatorien, die vor dieser Arbeit nicht in Europa zur Verfügung stand. Der entwickelte Prototyp kann bei verschiedensten Observatoriumskonzepten Anwendung finden und wird in den kommenden Jahren weiterentwickelt um allen Anforderungen einer zukünftigen Weltraummission gerecht zu werden.

Schlagworte: Gravitationswellen; Laserinterferometrie; Zeitrauschen

ACKNOWLEDGEMENTS

I wish to express my deep gratitude and sincere thanks to all who offered me their support, expertise and friendship. It was a joy working at the Albert Einstein Institute and the Department of Physics of the University of Florida in the past years. The professional environment and excellent work climate of these places are the merit of Prof. Dr. Karsten Danzmann and Prof. Dr. Guido Müller. This is to all the members of both institutes, the different research groups, IT departments, workshops, maintenance and administration, explicitly including everyone at the Centre for Quantum Engineering and Space-Time Research: thank you all! I am delighted by your friendship and hospitality and have greatly benefited from your assistance and expertise. Above all I am deeply indebted to my direct supervisor PD Dr. Gerhard Heinzel who supported me during all my research with passion and knowledge. Following closely behind are Dr. Michael Tröbs, Dr. Shawn Mitryk, Dr. Benjamin Sheard, and Dr. Yan Wang. Most of what I know about physics I learned from you.

International partners always provided much more support than anyone could expect. In particular I wish to mention: Oliver Jennrich from the European Space Research and Technology Centre for his customized orbit simulations; Anders Enggaard, Torben Rasmussen, and Torben Vendt Hansen from *Axcon ApS* for the many extremely valuable discussions and otherwise unobtainable hardware samples; Søren Møller Pedersen from the National Space Institute of Denmark for his great support and project management; and Sven Voigt from the *Northrop Grumman LITEF GmbH* for the construction and permanent loan of space qualified electro-optic modulators. Without your help and capacity for dialog my research would have been impossible. The same is true for the Max Planck Society, the Leibniz Universität Hannover, the Centre for Quantum Engineering and Space-Time Research, the University of Florida, and the European Space Agency who financed my research.

Beyond research, I was very lucky to be encouraged and supported in the development of my communication skills. I was given the opportunity to visit specialized training programs, help in public relations campaigns, and bring the fascination of my research to people all over the world. This was something I deeply enjoyed and that cannot be taken for granted. Over the years, an international circle of friends has formed that lasts beyond work. Speaking of friends: there are Gerhard Heinzel, Daniel Schütze, Marian Hövel, Andreas Wittchen, Fabian Zywietz, and Julian Barke, who—even though not obligated and undoubtedly tired of the subject—read through this thesis and helped to improve it. I owe you a great debt of gratitude! Finally, and maybe most importantly, I wish to thank my family who always supported me unconditionally. My parents awakened my interest in cosmology and enabled me to study at university. Together with my brother they provide a constant in my life that I can always count on. Thank you for your love and friendship! Without either of you it wouldn't make up a complete universe.

OVERVIEW

Chapter 1 is an introduction to gravity, cosmology, and the paradoxes that arise from a combination of general relativity and quantum physics. Gravitational waves can answer some of the most fundamental questions about our universe and allow us to study as yet unobservable objects. Different detection principles are presented. It becomes clear that laser interferometric gravitational wave observatories in space make the most rewarding range of frequencies accessible.

Chapter 2 explores the limits of such spaceborne observatories considering the detailed instrument design and the related noise sources. A web application is introduced that was developed to quickly explore the entire parameter space. This allows to assess many different mission concepts, to identify limiting influences, and to identify the most promising candidate within technological and budgetary constraints.

Chapter 3 presents a more detailed study required to determine the heterodyne frequency range for concepts of different arm lengths, laser relative intensity noise levels, and mission durations. The maximum heterodyne frequency affects the timing stability imposed on oscillators generating the local reference frequency for the beat-note measurements. In no case this requirement can be met with presently available technology.

Chapter 4 shows how to suppress the excess timing jitter by means of a universal reference signal. This signal needs to be generated, converted, and distributed locally and throughout the entire constellation by the Inter-Spacecraft Frequency Distribution System. This involves various electronics but also electro-optics and optical components.

Chapter 5 summarizes a technology assessment, development, and testing activity of Inter-Spacecraft Frequency Distribution System. It starts with the decision on the very basic principle of operation and results in the very first fully operational Frequency Distribution System. Over the cause of five years, devices were identified and hardware was built that complies with even the most demanding timing stability requirements.

Chapter 6 concludes this thesis and explains what has to be done to convert the current prototype of the Inter-Spacecraft Frequency Distribution System into viable flight hardware. A testbed that simulates the independent measurements on board the different spacecraft is currently under construction. It can be used to evaluate the entire metrology system under realistic conditions before the first gravitational wave observatory will launch into space.

CONTENTS

Laser source

Faraday isolator

Beam splitters
(cube and plate)

Mirrors
(normal and flipping)

Photoreceivers:

normal

dual output (HF/LF)

quadrant cell

Gravitational
reference sensor

Thruster

Optical fiber cable

Beam dump

Electro-optic phase modulator

Fiber amplifier

Waveform generators
(sine / PRN code)

Frequency converters
(multiplier / divider)

Filters
(high-pass / low-pass)

Adder

Power splitter
(power compiner)

Attenuator

Amplifier

PID controller

Mixer

Measurement
equipment

Directional
coupler

Bias tee

Rectifier

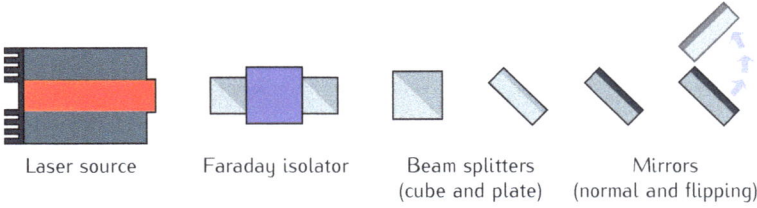

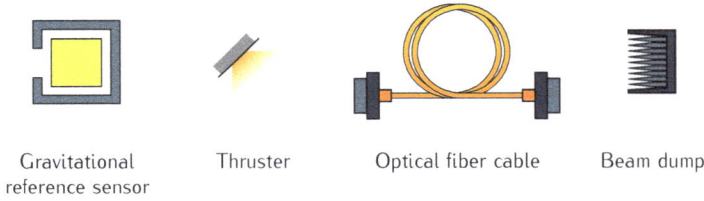

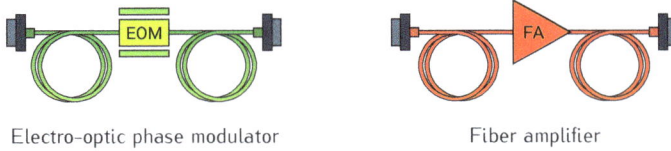

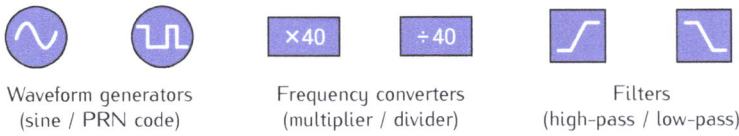

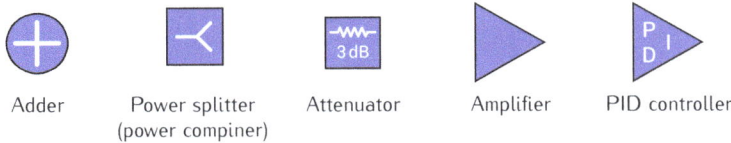

These icons were developed
as Symbol Library for
Adobe "Illustrator CC 2014"
and you may use them
under CC-BY (Creative
Commons Attribution
License).

simonbarke.com/phd/cl5

Part I

INTRODUCTION

Not even 400 years ago we punished scientists for sharing their findings with the world. Then there came Newton, Einstein, Hubble, and Hawking to open our eyes.

Currently we fly to planets, moons, and asteroids; we bring back samples and collect cosmic particles. Yet almost all our knowledge about the universe is based on the observation of electromagnetic waves—from radio waves over visible light to X-rays—detected on ground and in space. Electromagnetic radiation is the most successful messenger up to now, but it is unable to report on some of the most exciting phenomena of the universe. Black holes for example can only be observed indirectly, hence the processes of their formation and evolution are largely unknown. Due to the high density in the early universe, the first 400,000 years after the big bang are fundamentally obscured to our vision. Moreover, about 95% of all matter and energy—by name dark matter and dark energy—do not interact with electromagnetic radiation at all and are entirely invisible to us. But there is a way to overcome these fundamental limits.

After experimentally verifying almost all aspects of Einsteins field equations for decades, we are about to put our current understanding of gravity to work for us. Gravity is the dominating force in the universe and all known forms of matter and energy interact gravitationally. The continuous observation of gravitational waves will lead the way to amazing discoveries and will forever change our picture of the universe.

1

We have come a long way since Galileo Galilei was persecuted by the Catholic Church for his findings that the Earth revolves around the motionless Sun [1]. Yet people of all cultures and religions are still offended by actions and facts that contradict their belief system. Scientists on the other hand are happy to admit that what they know is not—and might never be—the final truth. They accrete knowledge through empirically observable results of reproducible experiments, which is known as the scientific method. Hence one of the fundamental differences between religious (or pseudoscientific) beliefs and scientific theories is the disposition of scientists to willingly enhance, replace, or discard an idea when new evidence contradicts the old findings. This was beautifully summarized by Albert Einstein.

> *No amount of experimentation can ever prove me right;*
> *a single experiment can prove me wrong.*
>
> — Albert Einstein [2]

Pseudoscience like homeopathy, astrology, or modern plasma cosmology is a claim or belief presented in a scientific way, but does not adhere to a valid scientific method and lacks supporting evidence.

Experimentation can only strengthen the likelihood that an idea is correct, but nothing can ever truly prove it. The strongest support comes if one can predict a result. This is how various hypotheses on relativity, space-time, and the relationship between mass and energy became established. Yet however elegant the idea, if nature is shown not to conform then the idea is wrong.

In fact, Galileo was wrong with his statement that the Sun is motionless. It rather orbits the center of our galaxy at a zippy 225 kilometers per second. Just recently NASA's Interstellar Boundary Explorer (IBEX) was able to map the solar system's tail of cosmic dust following behind [3]. Even our entire Milky Way galaxy is heading towards a collision with the Andromeda galaxy at roughly 110 kilometers per second [4], and our sun is dragged along. And then there is the 'Great Attractor', a gravity anomaly that attracts our entire Local Group of galaxies at roughly 600 kilometers per second [5]. All of these forces on our Solar System add up to a velocity relative to the cosmic microwave background – which is as close as we can get to a rest frame of the universe – of 371 km/s. So you see that Galileo's main finding about celestial mechanics is still valid: the Earth is by no means the center of anything. An impressive overview of systems bound by gravity can be found in Figure 1.1. Our own Milky Way galaxy is part of the Laniake Supercluster (see Figure 1.1.2). Traces represent the movement of galaxies in the direction of the Great Attractor.

When I write of a scientist "being wrong", you should read "being not entirely correct in every single detail".

The nature of this gigantic unseen mass some 250 million light years from our Solar System remains one of the great mysteries of astronomy.

Figure 1.1.3: NGC 4414, a typical spiral galaxy, contains 300 billion stars in a disc 100,000 light years across. *credit: NASA Headquarters (NASA-HQ-GRIN)*

Figure 1.1.2: The Laniakea Supercluster with 100,000 galaxies stretched out over 520 million light years. *credit: Nature art department, Mark A. Garlick*

Figure 1.1.1: The Cosmic Web, large scale structures of the universe on the scale of billions of light years. Bright spots represent superclusters of galaxies. *credit: Springel et al.*

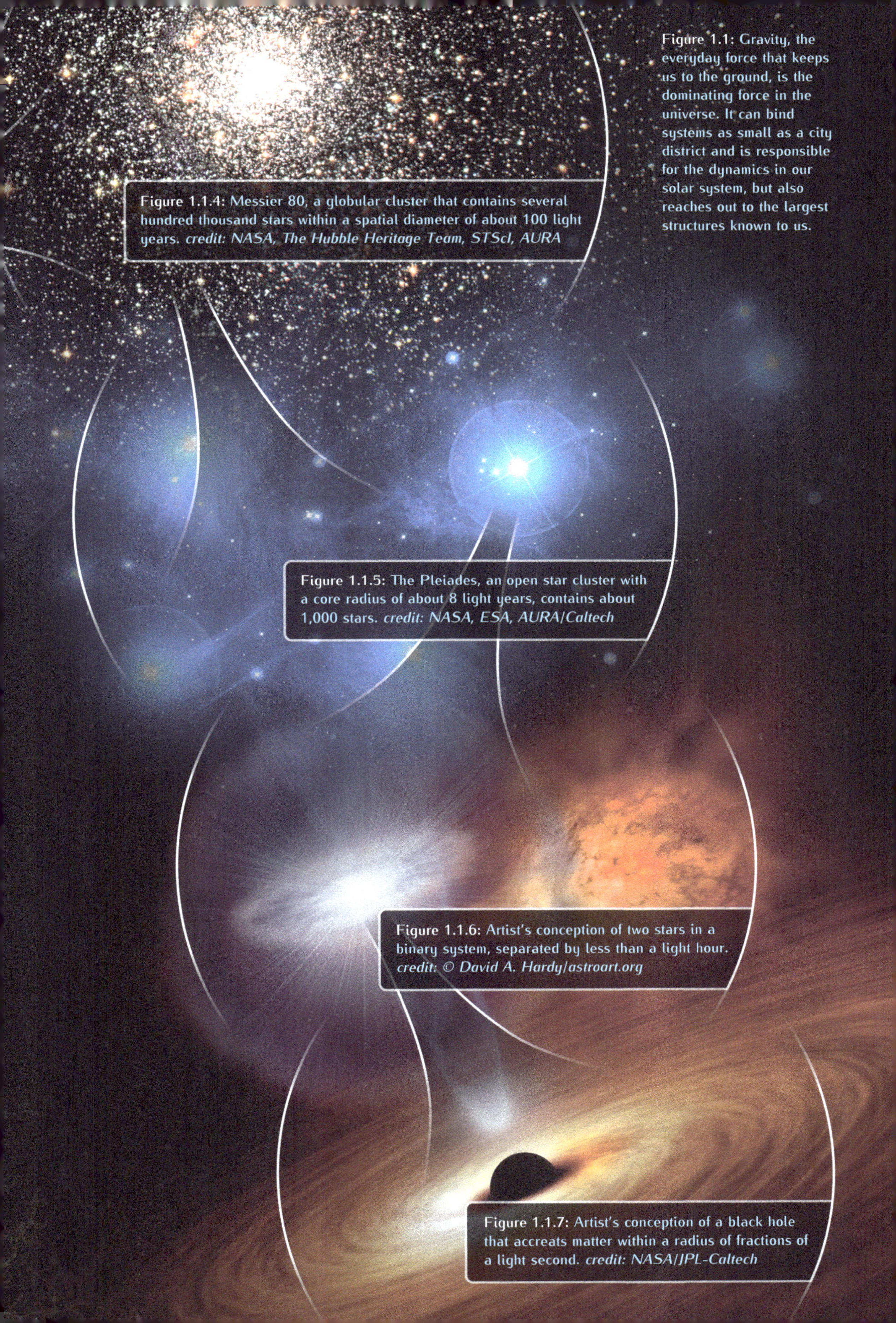

Figure 1.1: Gravity, the everyday force that keeps us to the ground, is the dominating force in the universe. It can bind systems as small as a city district and is responsible for the dynamics in our solar system, but also reaches out to the largest structures known to us.

Figure 1.1.4: Messier 80, a globular cluster that contains several hundred thousand stars within a spatial diameter of about 100 light years. *credit: NASA, The Hubble Heritage Team, STScI, AURA*

Figure 1.1.5: The Pleiades, an open star cluster with a core radius of about 8 light years, contains about 1,000 stars. *credit: NASA, ESA, AURA/Caltech*

Figure 1.1.6: Artist's conception of two stars in a binary system, separated by less than a light hour. *credit: © David A. Hardy/astroart.org*

Figure 1.1.7: Artist's conception of a black hole that accreats matter within a radius of fractions of a light second. *credit: NASA/JPL-Caltech*

In 1687 English physicist Sir Isaac Newton tried to explain the motions of moons and planets and found that two masses (m_1 and m_2) attract each other by a force (F) proportional to the product of the masses and inversely proportional to the square of the distance (r) between them:

$$F = G\frac{m_1 \times m_2}{r^2} . \tag{1}$$

The proportionality factor G in Newton's inverse-square law of gravity [6] is an empirical physical constant measured to be

$$G = 6.67384 \times 10^{-11}\,\mathrm{N(m/kg)^2} \tag{2}$$

with a relative standard uncertainty of 0.12%. All superclusters of galaxies, stellar associations, and planetary systems are bound by this invisible gravitational force that dominates all structures in the universe.

As you would expect from a good scientist, it was Newton himself who questioned his own theory. For him, the assumption that gravity acts instantaneously, regardless of distance and even through a vacuum, was "so great an absurdity that, I believe, no man who has in philosophic matters a competent faculty of thinking could ever fall into it." [7] Finally, a discrepancy in Mercury's orbit pointed out that Newton's theory must be wrong [8]. However, it provides a very accurate approximation and is still used today for most physical situations including calculations as critical as spacecraft trajectories.

Until the 19th century, many physicists tried to come up with a mechanical explanation of gravity without the troubling 'action at a distance'. Many of them included some kind of aether, a space-filling substance or field [9]. But all of these theories were overthrown by observations.

1.1 GENERAL RELATIVITY & THE UNIVERSE

Watch *Gravity Ink.* "Einstein's Gravity" for a quick introduction to the general theory of relativity.

simonbarke.com/phd/gi1

In 1907 Albert Einstein started working on a hypothesis as to the cause of the gravitational force. Nine years later he published his geometric description of gravity: the general theory of relativity [10].

His conclusion: gravity does not propagate through space, and it is not a force of a field, or substance penetrating empty space. Instead gravity is mediated by the deformation of spacetime, a mathematical model that combines the three dimensions of space and the one dimension of time into a single four-dimensional continuum. While mass, energy, momentum, pressure, or tension (all combined in the stress-energy tensor $T_{\mu\nu}$ which measures the matter content) curve spacetime, matter simply follows the geodesics of spacetime.

Spacetime tells matter how to move;
matter tells spacetime how to curve.

— John Wheeler [11]

This main message of general relativity is illustrated in Figure 1.2. The geometry of spacetime is described in the Einstein tensor $G_{\mu\nu}$, which measures its curvature. Each of the two tensors has 10 independent components, the relationship between both was formulated in the Einstein field equations as

$$G_{\mu\nu} + \Lambda g_{\mu\nu} = \frac{8\pi G}{c^4} T_{\mu\nu} \, , \tag{3}$$

where G is the same gravitational constant as in Newton's law (Equation 2) and c is the speed of light. This equation makes gravity a fictitious force where free falling reference frames are equivalent to an inertial reference frame. The Λ in Equation 3 is the cosmological constant, an energy density in otherwise empty space influencing the metric tensor $g_{\mu\nu}$ (or, simplified, the gravitational field).

1.1.1 THE SHAPE OF OUR UNIVERSE

In a static and never changing universe, as assumed by Einstein, a well chosen cosmological constant could counteract gravity and prevent the universe from falling in on itself. After astronomers like Georges Lemaître [13] or more famously Edwin Hubble [14] discovered that the universe is expanding and must have been created in a big bang from a 'primeval atom', it was not needed to artificially stabilize the universe any more, but: as it turns out, the same constant Λ can be used to describe the accelerated expansion observed in our universe [15]. As you see, the Einstein field equations are capable of much more than just explaining gravity; they have given us a tool set to understand the workings of the universe.

Lemaître derived Hubble's law and provided the first observational estimation of the Hubble constant in his original 1927 paper, but these parts were lost in translation for an English publication in 1931 [12].

When you assume a homogeneous and isotropic universe, you can derive a set of equations, called the Friedmann equations, that govern the expansion of space [16]. These equations tell you that a universe within the context of gravitational relativity could either be flat (i.e. Euclidean space), a closed

The assumption that the distribution of matter in the universe is **homogeneous**, also known as the *cosmological principle*, is justified on scales larger than 100 Mpc.

Figure 1.2: Illustration showing the Earth and the Moon warping the fabric of spacetime.

3-sphere of constant positive curvature, or an open 3-hyperboloid with a constant negative curvature [17]. As derived from one of the Friedmann equations, in a flat universe without a cosmological constant the mass density ρ would equate the critical density

$$\rho_c = \frac{3H^2}{8\pi G} \, . \tag{4}$$

Hence with no more than a given Hubble constant H (speed of expansion of the universe) and a gravitational constant G, you are able to conclude the shape, curvature, and fate of the universe out of its mass density. For $\rho < \rho_c$ the universe would be an open 3-hyperboloid and expand forever. For $\rho > \rho_c$ it would be a closed 3-sphere and eventually stop expanding, then collapse under its own gravity. It would also be of finite size and, due to its curvature, in the end traveling far enough in one direction will lead back to one's starting point. The special case of $\rho = \rho_c$ results in a flat (or Euclidean) and static universe as described above. These three cases are commonly expressed by the density parameter $\Omega \equiv \rho/\rho_c$ with $\Omega < 1$, $\Omega > 1$, and $\Omega = 1$ respectively, as illustrated in Figure 1.3.

It is important to note that there is a huge discrepancy between the baryonic matter density, ρ_b with $\Omega_b \equiv \rho_b/\rho_c$, and the mass density calculated through general relativistic means. The observed gravity within large-scale structures in the universe is much stronger than what could be accounted for by visible matter. Gravitational lensing—background radiation curved by gravitational fields—points to a total mass six times larger than what can be observed directly. Since free photons and cosmic neutrinos—that once accounted for big parts of the mass and energy distribution in the early universe—are entirely negligible nowadays, cosmologists hypothesized that this excess gravity is caused by an yet unknown form of 'dark' matter that does not interact electromagnetically. Current models assume that it consists of slowly moving particles which interact very weakly with electromagnetic radiation [18]. Thus these "cold dark matter" (CDM) particles are almost invisible and can currently only be observed through gravitational interaction. The CDM density, ρ_{cdm} with $\Omega_{cdm} \equiv \rho_{cdm}/\rho_c$, would also explain other

Baryonic matter accounts for all 'ordinary' matter and is usually referred to as visible or luminous matter.

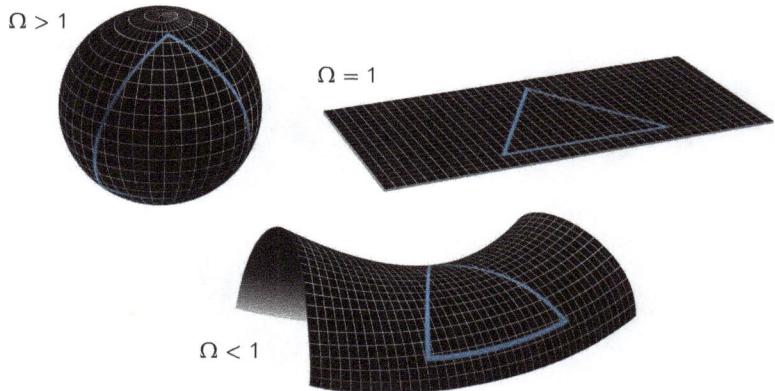

Figure 1.3: The geometry of the universe depends on the density parameter Ω: it is spherical for $\Omega > 1$, hyperbolic for $\Omega < 1$, and flat for $\Omega = 1$. The 3-dimensional structure of the universe is depicted as easily visualizable two-dimensional surfaces. *credit: licensed under public domain via Wikimedia Commons*

mysteries like the "flat" rotation curves of galaxies [19] or the evolution of large-scale structure of the universe [20].

Considering a positive cosmological constant (or vacuum energy density) the situation gets even more complex. Here you could have a closed, spherical universe where the vacuum energy density, ρ_Λ with $\Omega_\Lambda \equiv \rho_\Lambda/\rho_c$, is part of the total mass density. Usually gravity dominates in the long run and causes a spherical universe to contract eventually. In the case of $\Lambda > 0$ however, the negative pressure of the vacuum energy prevents that from happening and drives an accelerated expansion of the universe. Hence the expansion of such a universe will continue forever, up to a point where the observable part of the universe would be quite empty.

The concept of a field with negative pressure that accelerates the expansion of the universe is generally referred to as "dark energy", where the positive cosmological constant is just its simplest form. Cosmological models in which the universe contains such kind of a field are collectively subsumed under the heading Lambda-CDM (or ΛCDM) model. It is the current "standard model" of cosmology because of its precise agreement with observations. Of course, this model allows many different flavors of the universe to exist within the borders of Einstein's field equations. Naturally, scientists are eager to measure the parameters of our universe to determine its shape.

Today we know that our universe has a baryonic matter density of $\Omega_b = 0.0456 \pm 0.0016$ that is missing a CDM density of $\Omega_{cdm} = 0.227 \pm 0.014$ to account for all observed gravitational effect. Furthermore we see that the universe we live in expands with a Hubble constant of $H = 70.4^{+1.3}_{-1.4}$ km/s per megaparsec. High-precision measurements show that this expansion rate changes over time and reveal that the rate of expansion is accelerating from 7.5 billion years after the big bang onwards. We can conclude a vacuum energy density of $\Omega_\Lambda = 0.728^{+0.015}_{-0.016}$ and determine the age of the universe to be 13.75 ± 0.11 billion years [21].

It is most fascinating to see that—at least in our local observable universe—the total mass density equates the critical density as exactly as

$$\Omega = \Omega_b + \Omega_{cdm} + \Omega_\lambda = 1.0006^{+0.0306}_{-0.0316} . \tag{5}$$

This would mean that we live in a totally flat universe. The positive cosmological constant produces an ever accelerating expansion and points to a very unpleasant fate of the universe. Some time in the distant future the Hubble constant might become so large that even stars in galaxies are torn apart and the observation of distant stars would become physically impossible. The metric expansion of space might not even stop at atoms and subatomic particles, breaking all bonds in matter and creating a huge, dark, cold and empty universe [22].

The ultimate **fate of the** universe not only depends on the shape of the universe and the vacuum energy density, but also on the role vacuum energy will play as the universe ages. It is still unclear whether the total energy is conserved in general relativity as the universe expands.

1.1.2 THE NATURE OF THE UNIVERSE

All cosmology described above is solely based on the Einstein field equations. It turns out, our universe not only seems to be flat, but could actually have zero total energy [23]. This leads to the speculation that it may have been created in a coincidental quantum fluctuation. Such a closed system would not require any higher structure to provide a trigger mechanism for the big bang. Yet predictions of quantum mechanics seem to contradict predictions from general relativity, and suddenly the picture becomes much more complicated.

Black holes for example are a solution of the Einstein field equations and describe a massive singularity surrounded by a gravitational field so strong that – within a well-defined surface known as the event horizon – even light cannot escape. Nevertheless, the quantum field theory predicts that black holes evaporate over time due to quantum vacuum fluctuations (creation of particle-antiparticle pairs of virtual particles) at the black hole's event horizon. The escaping particle is known as Hawking radiation [25] and causes the black hole to lose mass and energy. It can be shown with currently accepted theories that this particle must be entangled with its infalling antiparticle that is swallowed by the black hole, as well as with all the Hawking radiation previously emitted by the black hole [26]. Since quantum mechanics forbids any particle to be fully entangled with two independent systems, the combination of general relativity, quantum mechanical unitarity, and quantum field theory creates a paradox [27].

The very same **Hawking radiation** might even prevent black holes from forming in the first place [24], yet we know from observations that black holes do exist.

There is another—much simpler—gedanken experiment that also tells us that the universe cannot be described by Einsteins field equations alone: the horizon problem. Naturally, we can only retrieve information from within a certain volume that is defined by the cosmological horizon which represents the boundary of the observable universe. Due to the nature of an expanding universe, this horizon has a radius of 46.2 billion light years [28] although light from that distance only traveled for 13.75 billion years (which represents the age of the universe). Light from outside this horizon had no chance to reach us yet. Thus it is obvious that we are embedded in a much larger unobservable structure that could have a different shape and where our local geometry only seems to be flat. Other parts of this larger structure, beyond our cosmological horizon, might host additional local universes of widely differing curvatures. Figure 1.4 shows Region A and Region B which both lie within our observable universe, but the local universes for both regions cover different parts of the overall universe and do not fully include each other.

These many **local universes** are very different to the parallel universes as a result of the many-worlds interpretation of quantum mechanics [29] where new universes pop into existence for every possible outcome whenever an observation is made.

This cosmological horizon imposes another paradox. We know that—due to its small size—the early universe was so dense and therefore so hot that photons scattered at free electrons. It took until 380,000 years after the big bang for protons and electrons to finally combine and form neutral hydrogen atoms. It was at this time that the universe became electromagnetically transparent. Today the light from this last scattering can be observed—now red shifted due to the cosmic expansion—in the cosmic microwave background (CMB) radiation. It tells us details about the conditions 380,000 years after the big bang.

Due to the random nature of the initial conditions the temperature of this radiation should be very different for different directions in the sky. Similar to Figure 1.4 distant regions in space back in the time of the last scattering had no causal contact. Thus there was no time to form an equilibrium and the initial temperature fluctuations should still be observable. Light sent out from opposite patches of the origin of the CMB just reached Earth, which is positioned half way between them. Thus those patches can not know anything about each other. Yet the CMB radiation has a surprisingly uniform temperature, isotropic to roughly one part in 100,000 over the entire sky, with a very fine fluctuation pattern that may have seeded the growth of structure in the universe. This cannot be explained by the standard ΛCDM model.

The resolution to this paradox has to be found somewhere before the time of last scattering. We can simulate the hot young universe from a second after the big bang and follow its evolution over time while it expands and cools down. The predictions of such simulations are at overwhelming agreement with detailed observations all the way to the present time. To explain the uniform CMB temperature physicists hence aim at the first fractions of a second after the big bang. Here, quantum effects become more and more important, but the theories we use to extrapolate back in time were not developed to include quantum physics. Until today, there is no "theory of everything" that describes gravity, space, time, and the shape and evolution of the universe, as well as physical phenomena at nanoscopic scales with all known quantum effects. It is impossible to tell if the properties of our universe gave birth to the laws of natural, or if there is an underlying fundamental set of rules that determines both, general relativity and quantum mechanics. Maybe the shape and nature of our universe is even somehow independent from the interactions of matter and energy within.

This is the earliest direct observation currently possible as we have no way to observe the universe prior to **that time**.

The distance between **Earth** and the origin of the CMB expaned to 46 billion light years. Thus even with a constant expansion of space at the current Hubble constant, light that just reached Earth could never reach the other side of the CMB: space in between Earth and the origin of the CMB expands so fast that both points seem to recede from each other faster than the speed of light.

The **first second after the big bang** is poorly understood in general. The unexplained imbalance in baryonic matter and antibaryonic matter for example originates from the same time.

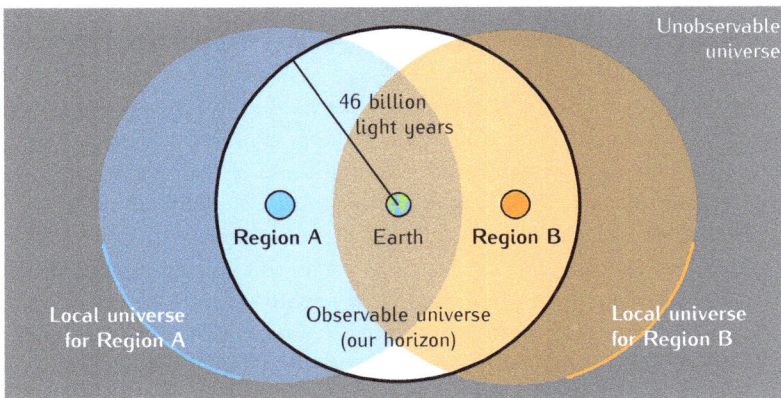

Figure 1.4: Horizons of local universes: many local universes can be embedded within a larger, unobservable structure. The horizon of these local universes is defined by the maximum distance from which one can retrieve information.

Currently two favored theories exist that extend the standard ΛCDM model to explain the observed uniform CMB temperature. The more conservative theory is the cosmic inflation model [30]. It postulates an inflationary epoch that lasted from 10^{-36} seconds to roughly 10^{-32} seconds after the Big Bang where space expanded exponentially. In this model, tiny quantum fluctuations during inflation became magnified to cosmic size, and all other inhomogeneities were smoothed out. It explains the uniform CMB temperatures including its pattern, and predicts that the total mass density equates the critical density, as can be observed today. In this model, in the beginning there was nothing. The universe started out of a singularity that expanded exponentially for a very short time span and follows the rules of general relativity ever since, including an ever accelerating expansion dominated by dark energy as shown in Figure 1.5. Inflation though was not the same for the whole universe – other parts would undergo different inflationary epochs or might even still inflate today, producing many local universes with very different properties.

Figure 1.5: The inflation model: The universe started out of a singularity, followed by a short epoch of inflationary expansion. As the universe cools down, its development is dominated by radiation, matter, and finally dark energy. Other parts of the universe would undergo different inflationary epochs, resulting in local universes with very different properties.

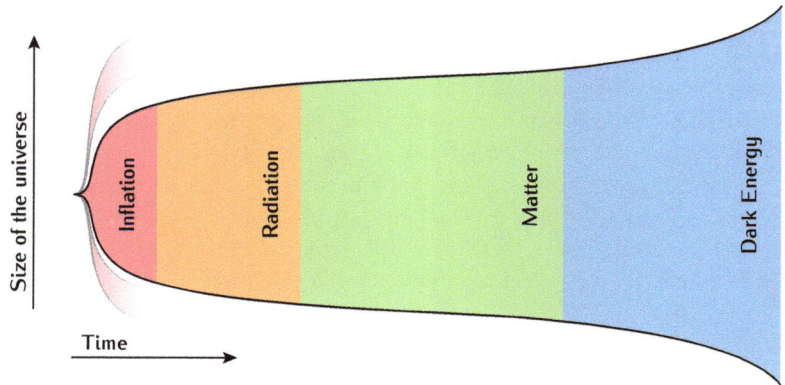

In M-theory—a unifying string theory and promising candidate for a unified theory of general relativity and quantum mechanics—a very different kind of universe is possible. This theory even raises the question if the big bang was the real beginning of our universe as it can also describe a cyclic (or 'ekpyrotic') universe that did not start with a singularity but with two branes moving apart [31]. Space in between both branes cannot be accessed by any object situated in either brane—like us—but both branes can move along this extra dimension. This opens the possibility for the big bang to be a collision of both branes, caused by a spring-like force between them [32]. In this collision hot matter and radiation was created. Both branes move apart while the two universes begin to expand. M-theory predicts that the accelerated expansion cannot last forever since within this context, dark energy is associated with the spring-like force between both branes. Eventually, it will bring both branes back together again. In the subsequent collision all kinetic energy is converted to new matter and radiation and the cycle starts all over again. This is illustrated in Figure 1.6.

Branes are higher-dimensional objects described by M-theory that exist in space-time and follow the rules of quantum mechanics

Strictly speaking this is not a full **cycle** as the universe contracts only in the extra dimension and continuously expands in all other dimensions. Thus the overall Universe becomes bigger and bigger. Nevertheless, the local universes from an observer's perspective remain the same for each cycle.

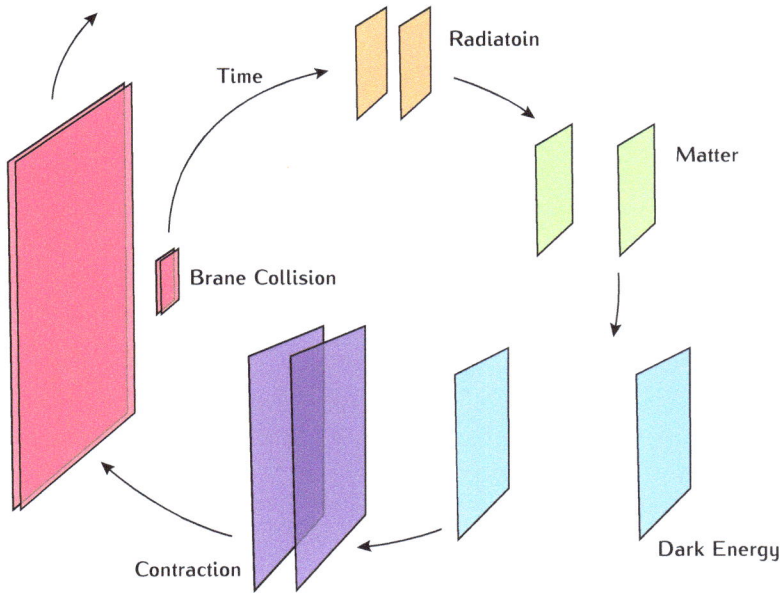

Figure 1.6: In M-theory, a cyclic model of the universe is possible: the collision of two branes generates hot matter and radiation, driving both expanding branes apart. Eventually, a spring-like force between both branes that is related to dark energy results in a contraction of both branes. Due to the ongoing expansion of the branes, the universe will be cold and empty just before both branes collide again. The collision itself generates new matter and radiation and the cycle starts over.

During the process of contraction – which is the alternative to inflation and lasts for about 10 billion years – the universe smooths out until quantum fluctuations take over. Consequently, the branes are slightly wrinkled and do not collide everywhere at the exact same time. Some regions of space bounce off earlier (heat up sooner) than others. Simulations predict that this would cause exactly the same uniform CMB temperature with its distinct pattern. This model also provides an explanation for the nature of dark matter: it merely is the influence of matter from the distant brane felt in our local universe.

Both models result in the universe that we observe today. The nature of the universe though is fundamentally different for both theories. In one, the CMB pattern is caused by quantum fluctuations shortly after the big bang during the time of cosmic inflation, while in the other one the same pattern arises from quantum fluctuations prior to the big bang which merely was the most recent collision of two higher-dimensional branes. The amount we already know about our universe speaks volumes of our ingenuity and science, but if we want to find answers about the origin of our universe, we need a new kind of observational cosmology. The only way to distinguish between both models lies in one single differentiation: strong gravitational waves should have been created during rapid inflation, but almost zero gravitational waves would originate from a slow collision of two branes. We just need to push todays technology a little bit further to detect these gravitational waves.

1.2 GRAVITATIONAL WAVES

Watch *Gravity Ink.* "The Future of Astronomy" for a quick introduction to **gravitational waves**.

simonbarke.com/phd/gi2

Besides black holes and gravitational lensing the maybe most exciting consequence of Einstein's theory is the postulation of **gravitational waves** [33]. The Einstein field equations predict that accelerated matter (or energy in general) emits gravitational quadrupole radiation as illustrated in Figure 1.7. These waves stretch and compress spacetime perpendicular to the direction of travel and cause directly observable distance fluctuations between freely falling objects. Let's assume we have a ring of cubes freely floating in the xy-plane and a gravitational wave propagates along the z-direction. As illustrated in Figure 1.8, the distance between the masses oscillates with time. The direction of this oscillation depends on the polarization of the gravitational wave. The usual basic set of polarization states are plus (+) and cross (×) polarization, others can be formed by linear combinations of these two.

While the strength of the gravitational field falls off with the square of the distance, this effect, an amplitude, falls off linearly proportional to the distance [34] and even sources located at the other end of the observable universe can produce relative distance fluctuations on the order of 10^{-20} or more, depending on the frequency of the signal.

EXAMPLE: On a distance of 4 kilometers, relative distance fluctuations of 10^{-20} correspond to 40 attometers. This is much smaller than an atomic nucleus. For a distance of 1 million kilometers, the same fluctuations correspond to 10 picometers, which is a factor of ten below the diameter of a hydrogen atom.

This effect, as tiny as it might seem, can tell us about electromagnetically invisible objects and has a huge discovery potential for new physics. Gravitational radiation travels unaffected throughout the entire universe; in contrast to electromagnetic radiation that interacts strongly with matter and hence can be distorted or blocked. Gravitational waves were able to propagate unimpeded even in the young hot universe prior to 380,000 years af-

Figure 1.7: Artist's impression of pulsar PSR J0348+0432 with white dwarf companion, finishing one orbit every 2.5 hours. This system is radiating gravitational waves. *credit: ESO / L. Calçada*

simonbarke.com/phd/gw

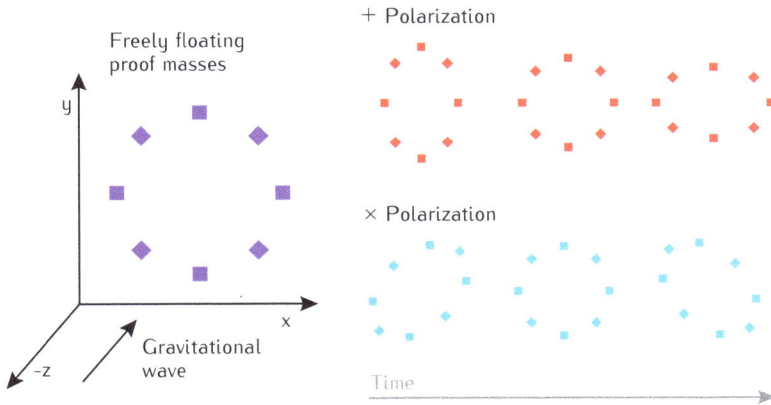

Figure 1.8: A ring of proof masses freely floating in the xy-plane and a gravitational wave that propagates along the z-direction. While a +-polarized wave will change the proper distance in x and y directions, the influence of a ×-polarized wave is rotated by 45° so that distances along the x- and y-axis remain unaffected.

simonbarke.com/phd/pol

ter the big bang. Thus gravitational waves are a superior messenger that holds complimentary or even otherwise completely unobtainable information about processes in the universe. Gravitational wave observatories are expected to bring the next big revelations in astronomy, cosmology, and fundamental physics alike.

Although general relativity passed all tests with flying colors, and despite indirect yet irrefutable proof of the existence of gravitational waves [35], gravitational waves have never been detected directly. Currently, research teams look into indirect evidence for gravitational waves produced during cosmic inflation, now red-shifted to a static polarization pattern imprint in the cosmic microwave background radiation [36, 37]. The clear detection or non-detection of a primordial gravitational wave background from the inflationary epoch would rule out one of the two leading theories about the origin of our universe. This—for me—is the most fascinating prospect of gravitational wave astronomy: we might not only determine the shape and structure of our own universe, but also learn about the nature of the global universe it is embedded in.

1.2.1   SOURCES

The primordial gravitational wave background should be observable over the entire frequency spectrum which makes a direct detection possible. Additionally there are many more sources that should produce gravitational waves. I grouped these sources into four main categories that determine the waveform of emitted gravitational waves.

Watch *Gravity Ink.* "The Gravitational Universe" for a quick introduction to gravitational wave detectors and **sources**.

simonbarke.com/phd/gi3

BURST SIGNALS   Rapid violent acceleration produces high frequency burst signals of distinct shape and characteristics. One known type of sources are Type II supernova events caused by the core collapse and rebounce of stars with mass greater than $\gtrsim 8 \ldots 10 \, M_\odot$ at the end of their thermonuclear burning life cycles (Figure 1.12.3) [38]. The strain amplitude depends on the asymmetry of these events. Observations of polariza-

tion in the spectra, jets in remnants, and kicks in neutron stars suggest that such supernovae are inherently aspherical [39]. Burst signals only last for a few milliseconds and thus lie in the frequency band between 100 and 1000 Hz.

CONTINUOUS SIGNALS Compact binary star systems or rapidly spinning neutron stars with smaller surface imperfections (on the order of centimeters) produce a stationary sinusoidal gravitational wave signal. Depending on the angular velocity of the orbit or rotation, this signal can be visible anywhere between sub-millihertz (Figure 1.12.5) and several kilohertz (Figure 1.12.6) frequencies. The radiation of gravitational waves will lead to a loss of angular momentum. A measurement of the orbital decay of the Hulse–Taylor binary pulsar (PSR B1913+16) is in total agreement with general relativity and was awarded the Nobel Prize in Physics in 1993 [40].

In particular, measurements **agree** with the expected energy loss due to the emission of gravitational waves.

INSPIRAL SIGNALS A sweeping sine signal occurs when two massive objects coalesce. This can be two supermassive black holes in the center of galaxies (Figure 1.12.2) that produce signals in the millihertz range—at amplitudes so high that it will be detectable throughout the entire visible universe—or a merging stellar mass binary system (Figure 1.12.7) producing a signal in the kilohertz range. When two objects with an extreme mass ratio inspiral (Figure 1.12.1), a complex waveform is generated by the highly relativistic orbit. Its harmonic frequencies and in particular their phase evolution over many cycles shed light on the detailed features of the spacetime in the close proximity to these objects. Additionally they tell us about their mass, spin and eccentricity at plunge, and will make it possible to distinguish between general relativity and alternative theories of gravity.

UNKNOWN SOURCES Since gravitational radiation is a completely new messenger never utilized before, it has enormous discovery potential. If the Big Bang and subsequent inflation caused a rapid expanding of spacetime, a gravitational wave echo dating from a period prior to the cosmic microwave background should exist (Figure 1.12.4).

The **Kardashev** scale is based on the amount of energy a civilization is able to utilize. Hypothetical Type II civilizations and higher are able to harvest the energy of their star or their entire galaxy. These civilizations are subject of current searches, e.g., in all-sky data of NASA's WISE telescope [41].

Other unknown sources like bursts of cosmic strings [42] are conceivable (Figure 1.12.8). Artificial sources of gravitational radiation from significantly advanced Kardashev Type II and III civilizations [43] manipulating energy equivalents of solar masses are theorized. Within the range of gravitational wave detectors, a lack of artificial signals would put upper limits on the existence of such civilizations.

This list is not intended to be exhaustive, but features a good overview of the variety of gravitational wave sources which is additionally summarized in Figure 1.12.

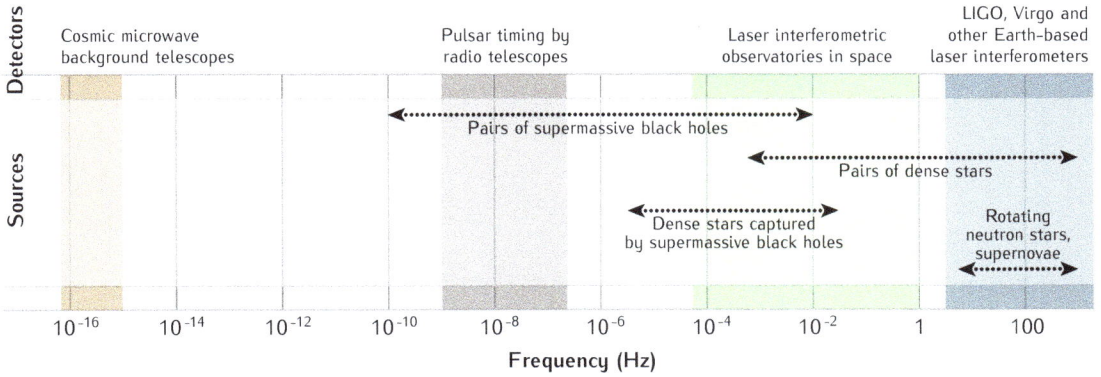

Figure 1.9: Frequency range of gravitational wave sources and bandwidth of corresponding gravitational wave detectors on Earth and in space. A gravitational wave background generated during cosmic inflation should be present over the entire frequency spectrum.

1.2.2 DETECTORS

The frequency components of gravitational waves depend on the nature of the source event and exist—not unlike electromagnetic radiations—in a wide frequency range from sub-millihertz up to the audio band. Naturally, there are a number of concepts on how to detect gravitational waves. For a quick glance of the frequency range of sources and detectors see Figure 1.9.

1.2.2.1 ON GROUND

Very low frequency gravitational waves below 10^{-6} Hz produced by pairs of supermassive black holes should be observable when measuring millisecond pulsar signals with ground based radio telescopes (e.g. the European Pulsar Timing Array [44]). Periodic shifts in these signals can be caused by gravitational waves. Since these detectors utilize the Earth-pulsar baseline of several kiloparsecs [45] they are sensitive for extremely long wavelengths (frequencies of $10^{-9} \ldots 10^{-6}$ Hz). Solid bars of metal (Weber bars) [46] on the other hand get excited at their resonant frequency by a matching gravitational wave passing by. The most promising concept for gravitational wave detectors at the kilohertz frequency range is a laser interferometric distance measurement. Using the Michelson topology as shown in Figure 1.10, a central laser is split in two directions (arms) and reflected by end mirrors back to the point of origin. Here both beams are superimposed.

These detectors are highly sensitive in a very **narrow frequency range** at a few 100 kHz and may detect gravitational waves produced by rotating neutron stars or asymmetric supernovae.

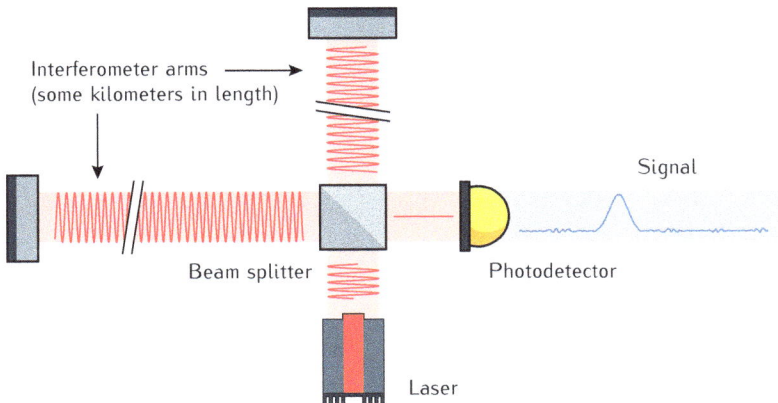

Figure 1.10: Michelson interferometer (homodyne detection) for gravitational wave detection. A laser is split and sent along km-scale arms. End mirrors reflect the beams back and the interference pattern is analyzed to detect relative distance fluctuations between the arms.

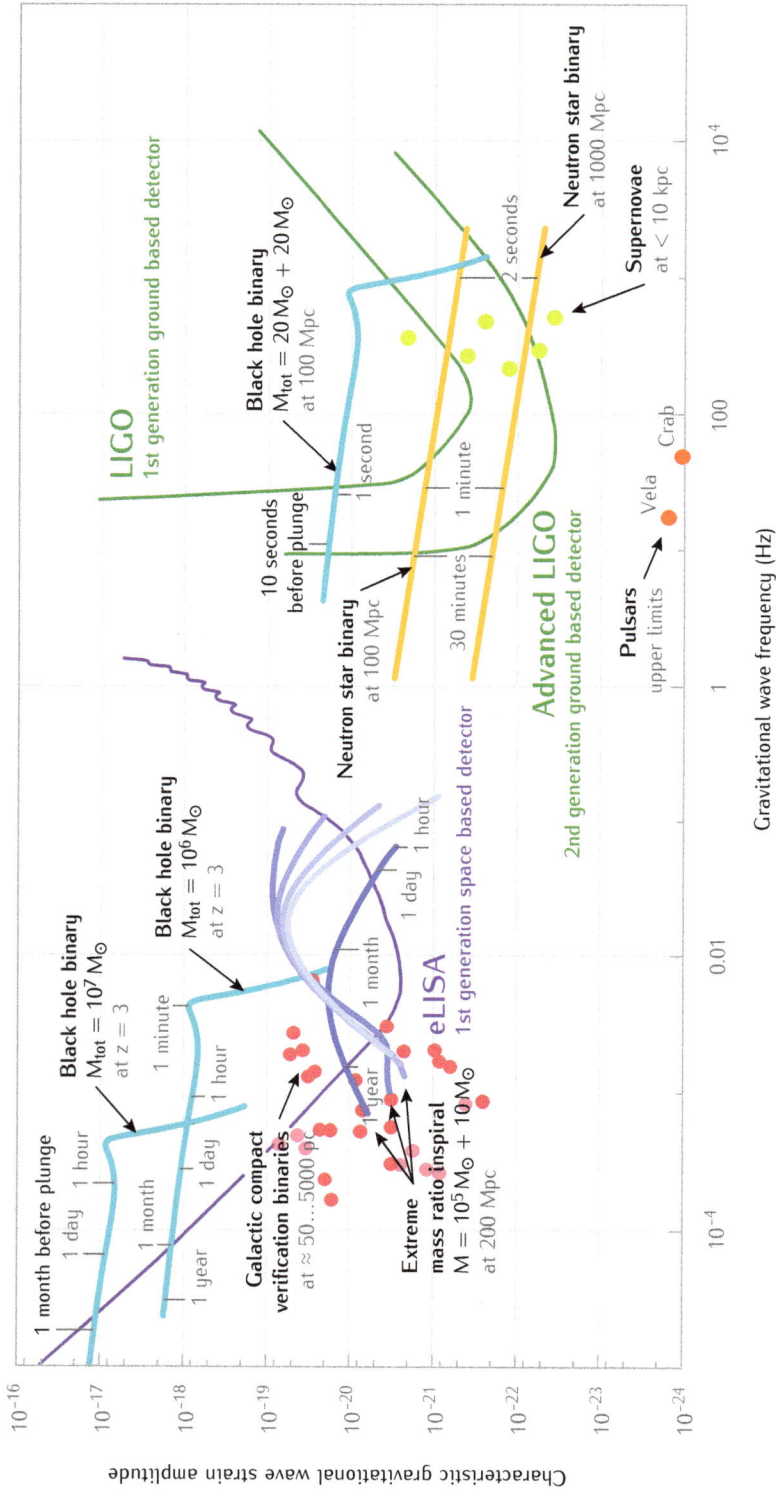

Figure 1.11: Characteristic gravitational wave strain amplitudes over signal frequency for a variety of signals (compare overview in Figure 1.12). Mass and distance of the events are indicated. The sensitivity of decommissioned 1st generation and upcoming 2nd generation ground based detectors is shown in green. They will detect stellar mass black hole binary mergers (Figure 1.12.2) and neutron star binary mergers (Figure 1.12.7), can set constrains on gravitational wave emissions of pulsars by matched filtering analysis (Figure 1.12.6) and may be able to detect burst signals of Type II supernovae (Figure 1.12.3). Designated space based detectors (as shown in blue) operate at much lower frequencies and will observe complementary sources of gravitational waves. They are sensitive to supermassive black hole binary mergers (Figure 1.12.2) within the entire observable universe, several thousand constant signals of Galactic compact binaries (Figure 1.12.5) some of which are already known, and the complex signal (harmonics shown in figure above) of extreme mass ratio inspirals (Figure 1.12.1) out to a distance of several hundreds of megaparsecs. The times before plunge are for inspiral and merger signals and indicate the maximum duration of observation within the given detector sensitivity.

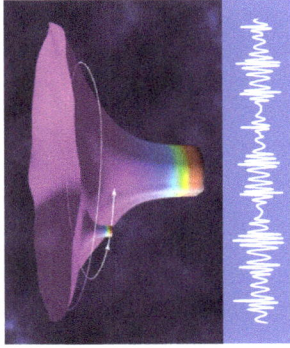

1.12.1: Extreme mass ratio inspirals. Neutron star or stellar mass black hole captured in a highly relativistic orbit around a massive black hole. Waveform: enormous amount of information. *credit: NASA*

1.12.2: Black hole binary mergers. Coalescence of two supermassive black holes (image: merging galaxy NGC 6240) or stellar mass black holes. Waveform: inspiral, merger, ringdown. *credit: NASA / ESA / HST / A. Evans*

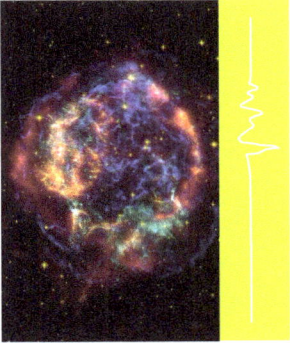

1.12.3: Spuernovae (Type II). Asymmetric core-collapse supernovae (image: Cassiopeia A remnant) produce gravitational waves. Waveform: distinct characteristics. *credit: NASA / ESA / JHU / R. Sankrit & W. Blair*

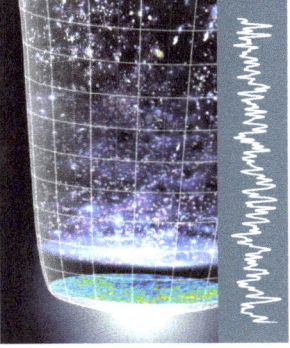

1.12.4: The Big Bang. Gravitational waves are the only form of information that can reach us directly from the Big Bang. Wave form: stochastic background (noise). *credit: NASA / WMAP science team*

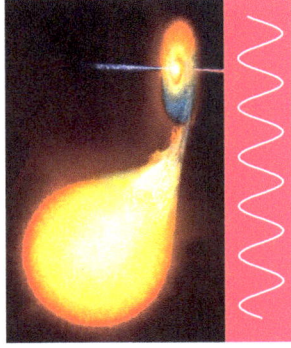

1.12.5: Compact binaries. A compact object like a neutron star or a black hole (right) and a companion in very close orbit. Matter transfer leads to Type Ia supernovae. Waveform: high frequency sine. *credit: ESA*

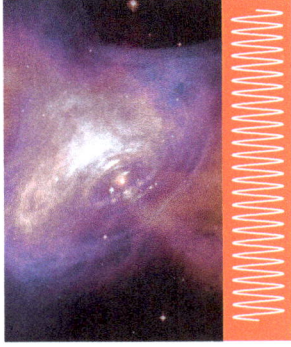

1.12.6: Pulsars. Rapidly spinning neutron star (image: Crab Pulsar) with small (sub-meter) surface imperfection. Waveform: high frequency sine. *credit: NASA / HST / CXC / ASU / J. Hester et al.*

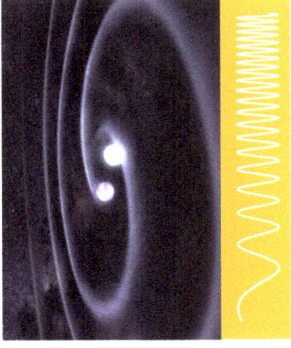

1.12.7: Neutron star mergers. Two neutron stars in a close orbit that shrinks until merger due to radiation of gravitational waves. Waveform: sine wave at increasing frequency. *credit: GSFC / D. Berry*

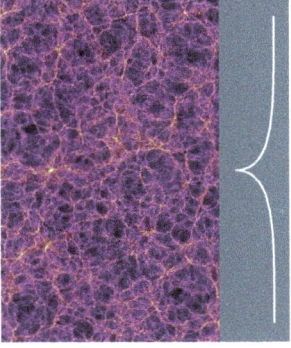

1.12.8: The Unknown. Gravitational waves can originate from cosmic string bursts or totally unknown phenomena invisible to electro-magnetic observations. *credit: MPI Astrophysics / V. Springel et al.*

Figure 1.12: Overview of different gravitational wave sources and corresponding typical waveforms of expected signals. Characteristic gravitational wave strain amplitudes, for example signals of inspirals (blue, cyan), burst events (lime) and constant signals (pink, orange, amber) are shown in Figure 1.11. The shape of a stochastic gravitational wave background or signals from yet unknown mechanisms (gray) are speculative.

The transition between constructive and destructive interference patterns (homodyne detection) reveal relative distance fluctuations between the beam splitter and the end mirrors that can be caused by gravitational waves. The measurement band is basically determined by the fixed arm lengths, but can be tuned with an additional resonant optical cavity ("signal recycling" or "resonant sideband extraction").

Current laser interferometric ground based detectors such as LIGO (USA), VIRGO (Italy and France) and GEO-600 (Germany and UK) [47–50] reach strain sensitivities of $2 \times 10^{-23} / \sqrt{\text{Hz}}$ at $\approx 200\,\text{Hz}$. The sensitivity of second generation detectors such as Advanced LIGO (USA and India) and KAGRA (Japan) is designed to improve that by one order of magnitude. They will be able to detect high frequency gravitational waves above $10\,\text{Hz}$ as produced by rotating neutron stars or asymmetric supernovae. However, all of these detectors are fundamentally limited to frequencies above $10\,\text{Hz}$ due to seismic disturbances and environmental gravity variations. This "seismic wall" is obvious in the LIGO and Advanced LIGO sensitivity plots (green traces in Figure 1.11, plotted in characteristic strain amplitude).

1.2.3 SPACEBORNE OBSERVATORIES

Due to these and other limitations of ground based detectors, the frequency range from $10\,\mu\text{Hz}$ to $10\,\text{Hz}$ might never be accessible from Earth. However, it contains some of the most exciting sources of gravitational waves and their observation may answer a wide range of fundamental questions, from the speed of gravitational interaction to the fate of the universe, as described in the white paper *The Gravitational Universe* [51]. It was recently selected as science theme by the European Space Agency (ESA) [52] and is publicly endorsed by renowned scientists such as Stephen Hawking and many Nobel and Fields Medal laureates [53]. To study the frequency range of interest, the white paper proposes a space based gravitational wave observatory, and ESA committed to a launch date in the 2030s as 3rd large mission of the Cosmic Vision program.

"LISA Mission" is the generic name for a laser interferometric gravitational detector concept which is the only credible candidate to answer The Gravitational Universe science theme selected by ESA for the L3 mission.

The only viable option for such an observatory known to date is a heterodyne laser interferometer with arm lengths of a few million km which was studied in great detail over the past decade. It became known as the LISA Mission. The sensitivity of one of the latest incarnations, eLISA, is designed to be $\approx 3 \times 10^{-20} / \sqrt{\text{Hz}}$ at $10\,\text{mHz}$ and plotted as dimensionless characteristic strain in Figure 1.11 (purple).

There are three categories of astrophysical phenomena that are known to emit gravitational waves at frequencies and amplitudes accessible to laser interferometric observatories in space.

1. **Massive black hole binaries**: the coalescence of two supermassive black holes (cyan traces in Figure 1.11).

simonbarke.com/phd/lisa

2. **Extreme Mass Ratio Inspirals (EMRIs)**: a compact star or stellar mass black hole captured in a highly relativistic orbit around a massive black hole (blue traces in Figure 1.11).

3. **Ultra-compact binaries**: systems of white dwarfs, neutron stars, or stellar mass black holes in tight orbit (pink circles in Figure 1.11).

The amount of energy emitted in form of gravitational waves is very different between these astrophysical phenomena. Thus the distance to detectable sources varies greatly.

MASSIVE BLACK HOLE BINARIES   Galaxies usually harbor one or more massive central black holes which are some million times heavier than our Sun. When galaxies coalesce, these black holes will merge eventually, releasing huge amounts of gravitational radiation during this process. Signals should be easily detectable out to redshifts of $z = 3$ and higher (at a distance of over ≈ 22 billion light years) even many months before the final plunge with increasing signal-to-noise ratios and predictabilities to the final event. Such gravitational waves originated over 12 billion years ago, so we can basically detect such events throughout the entire observable universe.

> The good **predictability** of the final plunge allows for a never-before-seen multimessenger astronomy with gravitational wave and electromagnetic observations. The early gravitational wave signals leave more than enough time to point other observatories to the region of interest.

Figure 1.11 shows two examples taken from [54]. In each case systems of two massive black holes at redshift of $z = 3$ are shown, one with a total mass $M_{tot} = 10^7\,M_\odot$, the other with $M_{tot} = 10^6\,M_\odot$. While the former signal starts at low frequencies approximately one month before the plunge (spike in the trace), the latter signal is shown for the final year before plunge. The detection of such signals will reveal the masses and spins of the two black holes, and shed light on the evolution and merger history of galaxies all the way back to shortly after the Big Bang.

EXTREME MASS RATIO INSPIRALS (EMRIS)   Compact stars or stellar mass black holes can be captured by the massive central black holes of galaxies. They are spiraling through the strongest gravitational field regions just a few Schwarzschild radii from the event horizon [51]. Such events should be resolvable many years before the merger for sources at hundreds of MPc distance. This corresponds to ≈ 2 billion light years and easily contains the entire Laniakea Supercluster and all neighboring structures, accumulating signals from over 500 million galaxies [55].

The highly relativistic orbits result in feature-rich waveforms with many harmonics. Figure 1.11 shows the first 5 harmonics of an eccentric EMRI for an object with mass $m = 10\,M_\odot$ captured by a massive black hole of mass $M = 10^5\,M_\odot$ at 200 Mpc distance [56]. The detection of such signals will allow a deep view into galactic nuclei for the very first time.

ULTRA-COMPACT BINARIES   About half of the stars in the Milky Way are thought to exist in binary systems [57], sometimes even in orbits so compact that orbital periods are shorter than one hour. A list of all currently known ultra-compact binaries can be found in [58]. For many of these systems, parameters (orbital period T, distance d, and masses m_1, m_2) are known with sufficient accuracy so we can calculate an order-of-magnitude gravitational wave signal prediction. Following [59] we find the dimensionless gravitational wave strain amplitude measured at a distance d from the source within one orbital frequency bin to be

$$h_c = 2 \left(4\pi\right)^{1/3} \times \frac{G^{5/3}}{c^4} f^{2/3} m \times M^{2/3} \times \frac{1}{d}, \qquad (6)$$

with $M = m_1 + m_2$ being the total mass and $m = \frac{m_1 \times m_2}{m_1 + m_2}$ the effective inertial mass. The frequency of the gravitational waves $f = 2 \times 1/T$ is twice the orbital frequency and G is the gravitational constant.

All known ultra-compact binaries are quasi-monochromatic meaning that they do not chirp appreciably during an observation of realistic length T_{obs}. Thus the frequency can be assumed to be constant over the mission duration $< T_{obs}$ and the signal amplitude accumulates to

$$h_c^{obs} = h_c \times \sqrt{N_{cycles}} . \qquad (7)$$

Here $N_{cycles} = f \times T_{obs}$ depicts the number of cycles observable within the observation time. Figure 1.11 shows all known ultra-compact binaries for $T_{obs} = 1$ year as pink circles.

We can observe Double white dwarf (WD) stars, ultra-compact X-ray binaries, AM Canum Venaticorum (AM CVn) stars, as well as any other cataclysmic variable (CV) stars, subdwarf B + WD binaries or double neutron stars out to distances of thousands of Pc. This corresponds to ≈ 30 thousand light years and includes our quadrant of the Milky Way galaxy with ≈ 50 billion stars. A small selection of known binaries is given in Table 1.

Table 1: A small selection of known double white dwarf (WD) stars, ultra-compact X-ray binaries, AM Canum Venaticorum (AM CVn) stars, and other cataclysmic variable (CV) stars with measured parameters. The resultant gravitational wave strain amplitude on Earth according to Equation 6 is given for 1 year of observation time. This is a rough estimate since the inclination of the orbital plane—which is not known for all of the binaries—was omitted. Detailed parameter inaccuracies can be found in [58].

Name	Type	Period T	Dist. d	Mass m_1	Mass m_1	Strain h_c
WZ Sge	CV star	4920 s	43 Pc	0.70 $M_\odot$	0.11 $M_\odot$	7.98×10^{-20}
HP Lib	AM CVn	1103 s	197 Pc	0.65 $M_\odot$	0.07 $M_\odot$	5.92×10^{-20}
SDSS J0651+2844	WD	765 s	1000 Pc	0.25 $M_\odot$	0.55 $M_\odot$	5.39×10^{-20}
WD 0957-666	WD	5270 s	135 Pc	0.37 $M_\odot$	0.32 $M_\odot$	3.80×10^{-20}
HM Cnc	AM CVn	322 s	5000 Pc	0.55 $M_\odot$	0.27 $M_\odot$	3.18×10^{-20}
EI Psc	CV star	3850 s	210 Pc	0.70 $M_\odot$	0.13 $M_\odot$	2.55×10^{-20}
SDSS J0923+3028	WD	3884 s	270 Pc	0.23 $M_\odot$	0.34 $M_\odot$	1.91×10^{-20}
SDSS J0926	AM CVn	1699 s	465 Pc	0.85 $M_\odot$	0.04 $M_\odot$	9.57×10^{-21}
SDSS J1436+5010	WD	3957 s	800 Pc	0.24 $M_\odot$	0.46 $M_\odot$	8.32×10^{-21}
4U 1820-30	X-ray	685 s	7600 Pc	1.40 $M_\odot$	0.06 $M_\odot$	4.04×10^{-21}
SDSS J0106-1000	WD	2346 s	2400 Pc	0.17 $M_\odot$	0.43 $M_\odot$	3.56×10^{-21}
4U 0513-40	X-ray	1020 s	12100 Pc	1.40 $M_\odot$	0.04 $M_\odot$	1.07×10^{-21}
XTE J1807-294	X-ray	2412 s	8500 Pc	1.40 $M_\odot$	0.02 $M_\odot$	2.80×10^{-22}

On top of that, there will be a noise contribution from the vast number of weak galactic binaries where individual sources cannot be disentangled in the data stream. The calculation of this noise usually involves a simulated catalog of millions of sources to find out how many sources are identifiable and which ones contribute to the overall noise floor, depending on the particular detector sensitivity.

This is just a description of the known sources accessible by one very special type of gravitational wave observatory. Above all, I wanted to make one point abundantly clear: gravitational wave detection is not an end in itself. It is not a matter of proving the existence of gravitational waves but about gaining knowledge about astronomy, cosmology, and fundamental physics—knowledge that is otherwise unaccessible.　　■

Part II

LIMITATIONS OF SPACEBORNE OBSERVATORIES

The first direct detection of gravitational waves will most likely happen within the next few years by laser interferometric ground based detectors. But even after this first detection we will continue to observe the universe with gravitational means—just as we still build telescopes, and still send probes and people into space. Event rates for ground based detectors are very limited. LISA-like space missions however as presented in the previous chapter cover a very rich frequency band of highly interesting sources. Thus the future clearly belongs to gravitational wave observatories in space.

It is of utmost importance to know the limiting factors of such spaceborne laser interferometric observatories. In Chapter 2 I will describe the main mission aspects, explain many detailed mission parameters and present all underlying calculations to derive a sensitivity curve as depicted in Figure 1.11. In the end, a web application to quickly explore the entire parameter space and design a realistic gravitational wave observatory is presented. Chapter 3 answers the more detailed questions about the laser interferometric heterodyne frequency range of current mission concepts. This topic has not been conclusively addressed as of this writing but introduces a number of mission requirements highly relevant for this thesis.

HOW TO DESIGN A GRAVITATIONAL WAVE OBSERVATORY

It is usually assumed that the noise floor of spaceborne gravitational wave observatories is dominated by optical shot noise in the signal readout. For this to be true, a careful balance of mission parameters is crucial to keep all other parasitic disturbances below the influence of shot noise. Based on previous work performed by members of the *Albert Einstein Institute* in Hanover, I developed a web application that uses over 30 input parameters and considers many important technical noise sources and noise suppression techniques to derive a realistic position noise budget. It optimizes free parameters automatically and generates a detailed report on all individual noise contributions. Thus one can easily explore the entire parameter space and design a realistic gravitational wave observatory.

The application is **based on work** by Gerhard Heinzel, Yan Wang, Juan Jose Esteban Delgado, Michael Tröbs, and the author himself [60].
It is publicly available at www.spacegravity.org.

In this chapter I describe the different parameters that are taken into account by the 'Gravitational Wave Observatory Designer', present all underlying calculations, and explain the final observatory's sensitivity curve.

spacegravity.org/designer

2.1 MISSION PARAMETERS

A laser interferometric gravitational wave observatory in space consists of a virtual Michelson interferometer that measures changes in the proper distance between gravitational reference points: freely floating proof masses that form the end mirrors of the interferometer arms. This concept is illustrated in Figure 2.1.

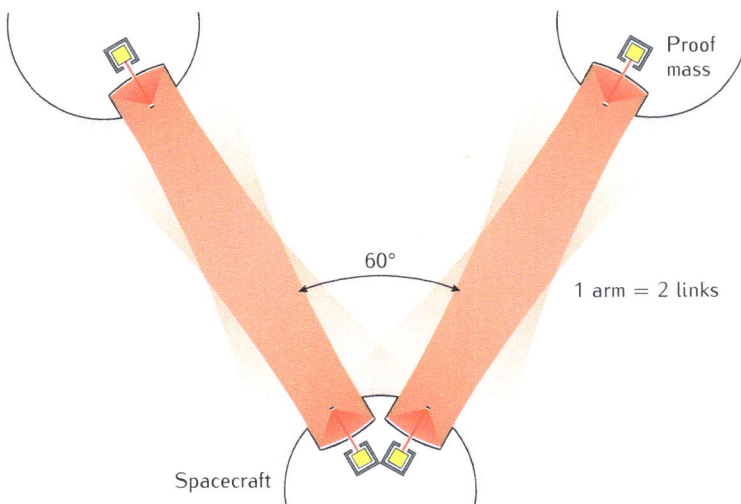

Gravitational waves will alter the **light travel time** in different proportions for the individual arms depending on their polarization and sky position.

60°

1 arm = 2 links

Proof mass

Spacecraft

Figure 2.1: A laser interferometric gravitational wave observatory in space consists of a minimum of three spacecraft that form a virtual two-arm Michelson interferometer with four individual laser links. Freely floating proof masses act as gravitational reference points.

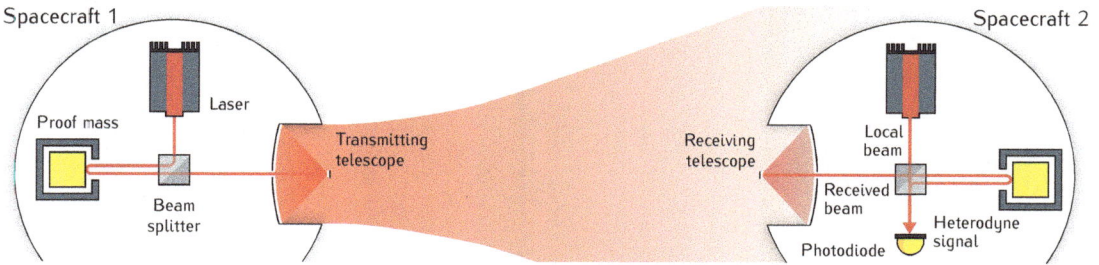

Figure 2.2: One laser link between two spacecraft of a gravitational wave observatory. A remote laser (on Spacecraft 1) is transmitted to Spacecraft 2. Here it gets interfered with a local laser of a different frequency and the heterodyne signal is detected. Gravitational waves alter the proper distance between the spacecraft resulting in a phase shift of the signal. Freely floating proof masses form the end points of the inter-spacecraft interferometer arm to suppress the influence of spacecraft position jitter on the actual arm length. To construct a complete observatory arm, one also needs the reverse link.

SAGA is a reference to Sága, a Norse goddess whose name translates to "seeress" in Old Norse language [61]. Sága is told to be a student of the universe, ever watchful and ever instructing us about the value of keen observation.

The virtual Michelson interferometer is constructed from individual 'links', each individual link consists of one or more actual laser interferometers, see Figure 2.2. Laser light from a distant spacecraft (received beam) is interfered with an on-board laser (local beam) at a recombination beam splitter on the local spacecraft. One observatory arm always consists of two counterpropagating links.

As an example, I will use a set of parameters that is currently assumed to be likely applied to the actual 2034 space mission by the European Space Agency. These parameters (all comprised in table Table 2) are quite different from previous concepts such as the 'Classic LISA' or 'eLISA (2013)' missions (see Section 3.1). Let this new gravitational wave observatory be known as 'SAGA' throughout this thesis for easier reference.

2.1.1 CONSTELLATION

There are a number of fundamental design choices that determine the capabilities of your observatory. While a minimum of two arms (four laser links) between three spacecraft is required to construct the virtual interferometer, more links will not only improve the observatory's sensitivity but also produce other consequential benefits: A triangular three-arm (6 link) detector can discriminate between different gravitational wave polarizations instantaneously and yields a much better spatial resolution. An octahedral 12-arm (24 link) observatory [62] would in theory be able to suppress acceleration noise on the proof masses alongside other else limiting noise sources. Possible arrangements are shown in Figure 2.3. For practical purposes, we only consider (nearly) equilateral constellations although other angles are feasible in principle.

Figure 2.3: Possible arrangements for interferometric gravitational wave observatories: two-arm (left), triangular (center), octahedral (right) – corner points mark the position of the individual spacecraft.

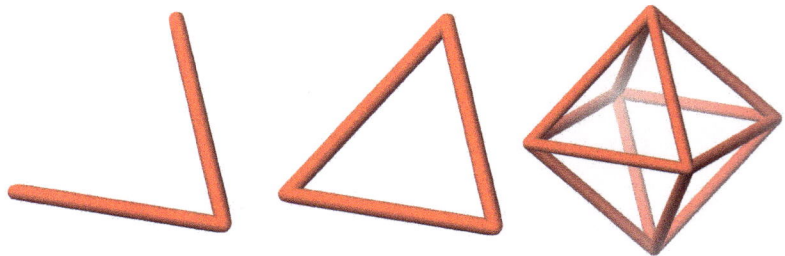

Parameter		Value
Number of links	$N_{\text{links}} =$	6
Average arm length	$L_{\text{arm}} =$	2 000 000 km
Heterodyne frequency (max.)	$f_{\text{het}} =$	18 MHz
Laser wavelength	$\lambda_{\text{laser}} =$	1064 nm
Optical power (to telescope)	$P_{\text{tel}} =$	1.65 W
Relative intensity noise (laser)	RIN $=$	$1 \times 10^{-8} / \sqrt{\text{Hz}}$
Laser frequency noise after pre-stabilization	$\widetilde{\nu}_{\text{pre}} =$	$290 \text{ Hz} / \sqrt{\text{Hz}}$
Telescope diameter	$d_{\text{tel}} =$	26 cm
Optical efficiency (receive path)	$\eta_{\text{opt}} =$	70 %
Beam waist position*		at transmitting telescope
Optimum beam waist radius*	$\omega_0 =$	11.60 cm
Received laser power*	$P_{\text{rec}} =$	585.62 pW
Local laser power*	$P_{\text{local}} =$	1.75×10^{-3} W
Temperature noise at electronics and electro-optics	$\widetilde{T}_{\text{el}}(f)$	see Section 2.1.4
Temperature noise at optical bench	$\widetilde{T}_{\text{ob}}(f)$	see Section 2.1.4
Photodiodes	$N_{\text{pd}} =$	4 segments
Quantum efficiency of photodiodes	$\eta_{\text{pd}} =$	80 %
Photodiode responsivity	$R_{\text{pd}} =$	0.69 A/W
Current noise (transimpedance amplifier)**	$\widetilde{I}_{\text{pd}} =$	$2 \text{ pA} / \sqrt{\text{Hz}}$
Capacitance (photodiode)**	$C_{\text{pd}} =$	10 pF
Voltage noise (transimpedance amplifier)**	$\widetilde{U}_{\text{pd}} =$	$2 \text{ nV} / \sqrt{\text{Hz}}$
Heterodyne efficiency	$\eta_{\text{het}} =$	70 %
Single first-order sideband power (in parts of carrier power)	$\frac{\text{sideband}}{\text{carrier}} =$	7.5 %
Modulation frequency	$f_{\text{mod}} =$	2.40 GHz
Timing jitter (electronics)	$\widetilde{t}_{\text{el}} =$	$4 \times 10^{-14} \text{ s} / \sqrt{\text{Hz}}$
Thermal stability (cables)	$\left(\frac{\delta\phi}{\delta T}\right)_{\text{cables}} =$	$7 \text{ mrad} / (\text{K m GHz})$
Thermal stability (fibers)	$\left(\frac{\delta\phi}{\delta T}\right)_{\text{fibers}} =$	$1 \text{ mrad} / (\text{K m GHz})$
Total length (cables)	$l_{\text{cables}} =$	2 m
Total length (fibers)	$l_{\text{fibers}} =$	5 m
Noise (EOM)	$\widetilde{x}_{\text{tml}}^{\text{eom}} =$	$3.81 \times 10^{-13} \text{ m} / \sqrt{\text{Hz}}$
Noise (fiber amplifier)	$\widetilde{x}_{\text{tml}}^{\text{fa}} =$	$7.62 \times 10^{-13} \text{ m} / \sqrt{\text{Hz}}$
Optical path length difference (in fused silica)	$\text{OPD}_{\text{fs}} =$	29 mm
Optical path length difference (on optical bench)	$\text{OPD}_{\text{ob}} =$	565 mm
Optical path length noise (telescope)	$\widetilde{x}_{\text{opn}}^{\text{tel}} =$	$1 \text{ pm} / \sqrt{\text{Hz}}$
Ranging accuracy (rms)***	$L_{\text{ranging}} =$	0.10 m
Acceleration noise	$\widetilde{x}_{\text{acc}}(f) =$	$3 \times 10^{-15} \frac{\text{m/s}^2}{\sqrt{\text{Hz}}} \times \frac{1}{(2\pi f)^2}$
Metrology system read-out noise	$\widetilde{x}_{\text{ms}}^{\text{pm}} =$	$1.02 \times 10^{-12} \text{ m} / \sqrt{\text{Hz}}$

Table 2: Parameters for the laser interferometric gravitational wave observatory 'SAGA'. These parameters are used to deduce the total equivalent displacement noise and observatory sensitivity. Many values were taken from a recent study by *Airbus Defence and Space* (formerly *EADS Astrium*) and are assumed to be likely applied to a space mission to be launched in 2034 by the European Space Agency. All parameters can be individually changed in the associated web application.

*Values were optimized automatically to achieve the lowest possible carrier read-out noise.
**These values contribute to the total photoreceiver equivalent input noise.
*** After raw data pre-processing.

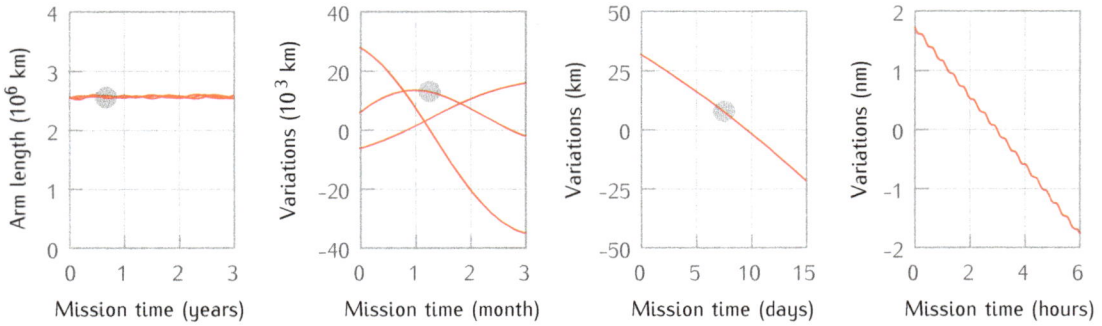

Figure 2.4: Exemplary inter-spacecraft distance (line-of-sight) for three spacecraft varies by many thousand kilometers over the cause of years and months (left). At days and hours this results in a quasi-constant drift superimposed by picometer variations due to the influence of gravitational waves at much higher frequency (right).

Additionally, there are many more possible arrangements. Even a single arm can detect gravitational waves but is omitted due to its very limited astrophysical capabilities and immense requirements on laser frequency and reference clock stability. Other arrangements like the Big Bang Observer (BBO) [63] consist of multiple three-arm observatories at different alignments and aims to improve the spatial resolution and polarization differentiation. These qualities are not yet addressed by the web application.

One of the most consequential mission parameters is the separation between spacecraft that determines the arm length of the virtual interferometer. It directly enters the conversion of displacement sensitivity to gravitational wave strain sensitivity [64, p. 74] and has additional multiple effects on the observatory's sensitivity. Longer arms make it more sensitive to lower gravitational wave frequencies but also decrease the received laser light power thus increasing the relative amount of shot noise in the signal. The gravitational wave sources commonly targeted by spaceborne observatories are in the millihertz range with wavelengths of 10^9 m and more, consequently the optimal arm length should be on the order of million kilometers. Even the observation of gravitational waves at hertz with cycle durations of the order of seconds still favors arm lengths of some thousand kilometers.

Ground based detectors are designed to have constant and well-balanced arm lengths so that laser frequency noise automatically cancels and homodyne detection can be used. In space, however, the inter spacecraft separation distance varies over the duration of the mission due to gravitational influences by the Earth–Moon system (and other planets in the solar system) on the individual spacecraft orbits. This so-called constellation or arm length "breathing"—which happens at much lower frequencies than visible influences by gravitational waves as illustrated in Figure 2.4—results in a relative velocity in the line-of-sight.

The line-of-sight velocity translates to a shift Δf in the frequency $f = c/1064\,\text{nm}$ of the received laser light (optical Doppler effect) according to

$$\Delta f = f \left(\frac{\sqrt{c + \Delta v}}{\sqrt{c - \Delta v}} - 1 \right) . \tag{8}$$

This prevents the use of homodyne read-out schemes and we have to deal with the interference of two laser beams of unequal frequency.

Homodyne detection describes interference between two beams at the same frequency. Because of constructive and destructive interference a phase difference between the two beams results in a change in the intensity of the light on the detector.

2.1.2 HETERODYNE INTERFEROMETRY

In our case, we have to assume different frequencies for the light from the incoming laser (f_{RX}) and the local laser (f_{LO}) with amplitudes E_{RX} and E_{LO}, and electric fields given by

$$E_{RX} \sin\left(2\pi f_{RX}t\right), \quad E_{LO} \sin\left(2\pi f_{LO}t\right).$$

Both fields interfere at a beam splitter and the light is detected by a photodiode as illustrated in Figure 2.5. The interference creates two new signals at the sum $f_{RX} + f_{LO}$ and the difference $f_{RX} - f_{LO}$ frequencies with intensities

$$\propto E_{RX}E_{LO}\left[\cos\left(2\pi\left[f_{RX} - f_{LO}\right]t\right) - \cos\left(2\pi\left[f_{RX} + f_{LO}\right]t\right)\right].$$

Due to the limited bandwidth of the photodiode, only the beat note at the difference frequency is detectable which is called heterodyne signal. The optical output signal will have an amplitude proportional to the product of the amplitudes of the input light fields. Within the linear range of the photodiode, it can be assumed that the output current of the photodiode is proportional to the optical signal intensity and hence to the squared amplitude of the electrical field.

The **intensity** of the down-mixed difference frequency can (and will) be larger than the intensity of the incoming laser light itself. This is because the weak incoming light is mixed with the much stronger local laser source.

In the presence of a gravitational wave which acts as a phase modulation $\phi(t)$ on the incoming laser light, this phase modulation will be conserved:

$$E_{RX} \sin\left(2\pi f_{RX}t + \phi(t)\right) \quad \times \quad E_{LO} \sin\left(2\pi f_{LO}t\right)$$
$$\Rightarrow \quad E_{RX}E_{LO} \cos\left(2\pi\left[f_{RX} - f_{LO}\right]t - \phi(t)\right).$$

If the optical phase of the incoming beam shifts by a certain phase angle, then the phase of the heterodyne signal shifts by exactly the same angle. Thus the phase of the heterodyne signal contains information of the the gravitational wave signal. This correspondence is illustrated in Figure 2.6.

The **picometer scale** variation of the arm length caused by gravitational waves can be separated from large motions due to inter-spacecraft drifts since they happen at different timescales (minutes vs. month).

EXAMPLE: A strong gravitational wave changes the spacecraft separation by 100 pm. This shifts the phase of the laser light (at 1064 nm wavelength) by 100 millionth of a cycle, or $\approx 600\,\mu\text{rad}$. The phase shift

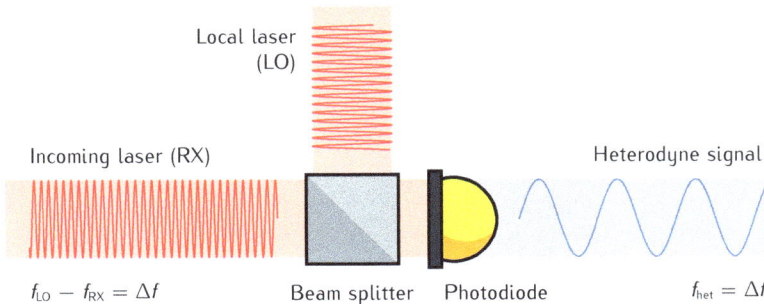

Local laser (LO)

Incoming laser (RX)

$f_{LO} - f_{RX} = \Delta f$

Beam splitter   Photodiode

Heterodyne signal

$f_{het} = \Delta f$

Figure 2.5: LISA uses heterodyne interferometry. The incoming laser light is combined with a local laser at similar (but not equal) frequency. A photodiode detects the heterodyne signal at the difference frequency and converts it to an electrical signal.

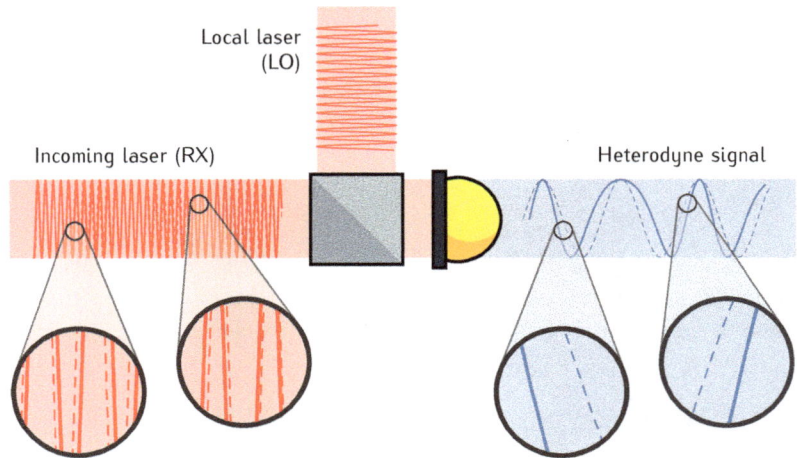

Figure 2.6: In the presence of a gravitational wave the incoming laser light is phase modulated. This phase modulation (solid vs dashed lines) is conserved in the heterodyne signal which makes it much easier to detect it at low frequencies. Please note that in reality the phase modulation will only be of μrad amplitude at mHz frequencies while the heterodyne signal is at MHz frequencies.

is conserved in the heterodyne signal. For a 10 MHz difference frequency (wavelength of $\approx 30\,\mu m$) the same $600\,\mu rad$ now correspond to a length of $\approx 3\,mm$ in the read-out. This is equivalent to a 10^{-11} seconds shift in the signal's arrival time compared to a much harder to detect 3×10^{-19} seconds shift in the original signal.

While other orbit parameters influence many different aspects of the observatory such as lifetime, solar power, and data transfer, the **maximum heterodyne frequency** currently is the only orbit-related aspect considered by the web application.

The phase read-out becomes easier with smaller heterodyne frequency. Hence it is desirable to keep all heterodyne frequencies as low as possible. This – and the necessity to avoid the case of zero heterodyne frequencies and other forbidden frequency domains – requires the implementation of adaptable offset frequencies (offset frequency phase-locked loops) between the different lasers in combination with a sophisticated frequency plan. The effectiveness of this effort is limited by the combined magnitude of inter-spacecraft drifts in a certain locking scheme.

The general feasibility of a chosen constellation with a specific spacecraft separation in a certain orbit is subject to a more detailed study in Chapter 3 that addresses forbidden frequency domains and the time-varying Doppler shifts in detail. For a customized laser locking scheme and frequency swapping plan that considers a wide variety of auxiliary functions [65] and technical limitations I was able to determine the maximum heterodyne frequency for a number of different constellations and arm length. In the following we will work with one specific case, a triangular three-arm ($N_{links} = 6$ links) formation featuring a reasonable arm length of $L_{arm} = 2\,000\,000\,km$ in a heliocentric orbit that results in a maximum heterodyne frequency of $f_{het} = 18\,MHz$.

2.1.3 LASERS, OPTICS, AND PHOTORECEIVERS

To decrease the read-out noise level of the observatory under investigation, it is not only beneficial to have high-quality photoreceivers but also to increase and stabilize the laser power received by the remote spacecraft (see

Section 2.2.1). For the power increase one can shrink down the arm length (which has an adverse effect on the overall sensitivity) or increase the laser power and enlarge the optical telescopes. The latter would result in a higher power consumption and a larger spacecraft and thereby a higher mission cost. Balancing these parameters within the mission's financial constraints is crucial.

LASERS   All lasers have to meet certain stability requirements. Fluctuations in the laser power relative to the average absolute power level, the so-called relative intensity noise (RIN), will directly couple to the photocurrent of the receiving photo detector as one part of the read-out noise and deteriorate the interferometric length measurements. The best space qualified lasers available as of this writing meet a relative intensity noise of $RIN = 1 \times 10^{-8}/\sqrt{\mathrm{Hz}}$ for Fourier frequencies above $5\,\mathrm{MHz}$ at $\lambda_{\mathrm{laser}} = 1064\,\mathrm{nm}$ wavelength [66]. Below this frequency the noise increases significantly so that no measurements at heterodyne frequencies below $5\,\mathrm{MHz}$ are possible. This limitation determines a forbidden domain for the frequency swapping plan that is described in detail in Section 3.2 and in turn determines the maximum heterodyne frequency. For other relative intensity noise levels this lower frequency might be different.

Frequency noise of the lasers will couple via the arm length difference of individual interferometers into phase fluctuations in the signal read-out. That is why one master laser is pre-stabilized by a reference cavity, a molecular frequency standard or similar techniques [67, 68], and all other lasers will be actively locked onto this master laser. The residual frequency noise after pre-stabilization is assumed to be $\tilde{\nu}_{\mathrm{pre}} = 290\,\mathrm{Hz}/\sqrt{\mathrm{Hz}}$. To simplify calculations and slim down the user interface of the web application, this noise contribution—like most within in this chapter—is given as white noise valid at the targeted gravitational wave frequency range.

The laser power—or, more importantly, the power passed to the transmitting telescope—possibly depends not only on the actual master laser but also on a laser amplifier. The above values for relative intensity noise and frequency noise after pre-stabilization already consider the presence of such an amplifier stage. In the following we consider a power passed to the transmitting telescope $P_{\mathrm{tel}} = 1.65\,\mathrm{W}$.

RECEIVED LASER POWER   Imperfections in beam pointing, a property that is potentially sensitivity limiting, is currently not considered by the web application. For the amount of light transmitted between spacecraft, only the telescope diameter and arm length parameters are used. In the following, we will assume a telescope with a moderate $d_{\mathrm{tel}} = 26\,\mathrm{cm}$ diameter primary mirror. We can now calculate the laser power received by the remote spacecraft. There are three different cases.

1064 nm is a standard wavelength for gravitational wave observatories. At other wavelengths relative intensity noise and frequency noise might be very different. There are additional consequences: While phase noise would have a smaller impact on the displacement noise at shorter wavelength (see Equation 14), drifts between spacecraft would result in higher Doppler shifts and hence increase the maximum heterodyne frequency.

Future versions of the web application will feature an advanced **user interface** for a more detailed frequency dependent description of individual noise sources.

1. **Short arms / big telescope mirrors**, where the full Gaussian beam fits well within the telescope when the waist is located at the center between the spacecraft. Here we can transmit the full laser power.

2. **Long arms / small telescope mirrors**, where the Gaussian beam has expanded to a width much larger than the receiving telescope when the waist is located at the telescope aperture. Here we cut out a 'flat-top' beam out of a field of constant intensity.

3. **Anything in between**, where the Gaussian beam is larger than the telescope diameter but too small for a flat intensity profile. This case is currently not addressed by the web application. It should hence be avoided since the received power will be highly affected by beam pointing which is a potentially limiting noise source.

To check if we can transmit the full laser power by setting the waist of the beam at the center between the spacecraft separated by L_{arm}, we compute the optimum waist radius ω_0 for a minimum Gaussian beam radius $\omega(x)$ at $x = L_{\text{arm}}/2$ apart from the waist:

$$\omega(x) = \omega_0 \times \sqrt{1 + \left(\frac{x \times \lambda_{\text{laser}}}{\pi \omega_0^2} \right)^2}. \tag{9}$$

For an arm length of $2\,000\,000$ km, the optimum waist is found to be 18.40 m and the observatory would require telescopes with a diameter larger than 50 m to transmit the full laser power. Thus we abandon this plan and intend to optimize the beam parameters for a maximum light intensity across the receiving telescope. As deduced from [64] the maximum intensity is reached by placing the waist at the transmitting telescope's aperture. For long arms the on-axis far-field intensity at the receiver can then be expressed as

$$I_{\text{rec}} = \frac{\pi \, P_{\text{tel}} \, d_{\text{tel}}^2}{2 \, L_{\text{arm}}^2 \, \lambda_{\text{laser}}^2} \times \underbrace{\alpha^2 e^{-\frac{2}{\alpha^2}} \left(e^{\frac{1}{\alpha^2}} - 1 \right)^2}_{\text{max}()=0.4073 \text{ for } \alpha=0.8921}, \tag{10}$$

where α is the waist radius in units of the telescope radius: $\omega_0 = \alpha \times d_{\text{tel}}/2$. The maximum of this function occurs at $\alpha = 0.8921$ as indicated above, so that the optimum waist radius is $\omega_0 = 0.8921 \times d_{\text{tel}}/2 = 11.6$ cm. Accordingly, the best achievable intensity at the receiver is $I_{\text{rec}} = 15.76\,\text{nW}/\text{m}^2$.

If we use a smaller beam that completely passes through the telescope, its divergence would be larger and the beam would be spread over a bigger area at the receiver. In consequence, the intensity would be smaller. If we use a larger beam with a smaller divergence, a larger fraction of the beam power would be rejected by the transmitting telescope aperture and again the intensity at the receiver would be smaller. In the equation above, diffraction effects for the beam truncated by a circular aperture were taken into account. The off-axis intensity distribution shows some curvature and diffraction rings, so that strictly speaking one cannot state a Gaussian beam radius. However, following Equation 9 to get an approximate far end beam diameter, we obtain $d_{rec} = 2 \times \omega(L_{arm}) = 11.68\,km$. This is much larger than the telescope diameter and we can confidently assume a flat intensity profile.

The received laser power now easily results from the light intensity at the receiving telescope multiplied with its optical efficiency and the collection area,

$$P_{rec} = \pi \left(\frac{d_{tel}}{2}\right)^2 \eta_{opt} I_{rec} = 585.62\,pW\,. \tag{11}$$

Here $\eta_{opt} = 70\%$ denotes an overall optical efficiency in the receive path that accounts for all losses in the optical path from the transmitting telescope to the recombination beam splitter on the receiving spacecraft.

OPTICAL BENCH   Interferometers are used to optically read out the displacement of the proof masses. These interferometers are typically constructed with fused silica optics that are bonded to an optical bench made out of an ultra-low expansion glass-ceramic [69]. There are different possible interferometer topologies. In principle the simplistic scheme illustrated in Figure 2.2 would suffice since the difference of the two heterodyne signals (both links) cancels not only noise induced by the laser feeds (optical fibers from the laser to the optical bench) but also spacecraft position noise and even the phase noise caused by temperature fluctuations of the optical bench. At the same time changes in the proper distance between the spacecraft (including gravitational waves) are preserved.

More complex topologies exist that split the single link measurement into smaller sections that are read out by individual interferometers [70]. For example one could omit the reflection of the received beam on the local proof mass. Instead, the proof mass displacement would then be determined with respect to the optical bench with a dedicated proof mass interferometer. This simplifies integration and testing of the interferometers and allows for eas-

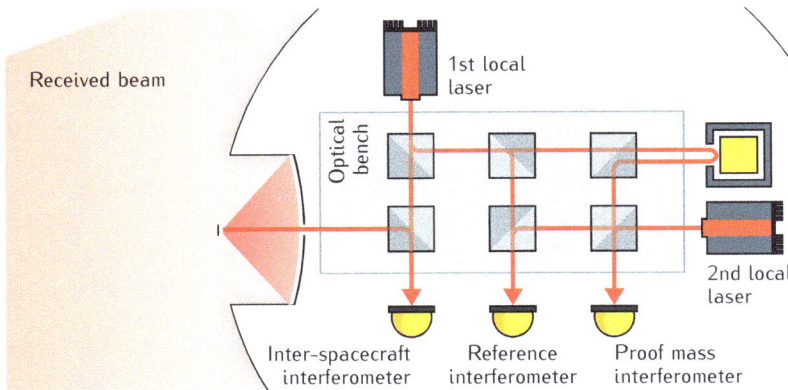

Figure 2.7: The measurement of the proper distance between any two proof masses is split into individual interferometers. Here, at each end of the link there are three interferometers, one to read out the inter-spacecraft distance, one to determine the displacement of the local proof mass in relation to the optical bench, and one acting as a reference.

ier beam alignment. Observatories that receive only low optical power from the remote spacecraft benefit from a scheme with three interferometers as illustrated in Figure 2.7. Here, a second local laser is used in the proof mass interferometer so that the full power of the received beam can be utilized in an inter-spacecraft interferometer. This scheme requires an additional reference interferometer to cancel the noise induced by the laser feeds. The web application is not limited to one certain read-out scheme. Usually it is assumed that a three interferometer scheme is used. However, with a proper change of the optical efficiency in the receive path (Equation 11) and optical path length difference (Section 2.2.3) parameters one can account for different schemes as well.

The heterodyne signal with the lowest amplitude (and thus possibly a limiting factor) usually is the one of the inter-spacecraft interferometer. Here, the heterodyne efficiency at the recombination beam splitter—a factor describing the mode overlap between the two laser beams—gains importance. It is assumed to be $\eta_{\text{het}} = 70\%$ which represents an estimate that results from parameters of the two interfering beams such as beam diameter, wavefront curvatures, and wavefront errors. A higher efficiency increases the signal that is received by the photo detector.

PHOTORECEIVERS   The heterodyne signal of the beam interfered at the recombination beam splitter is detected by a photodiode. A transimpedance amplifier converts the photocurrent into a proportional voltage. The quantum efficiency of the photo detector is assumed to be $\eta_{\text{pd}} = 80\%$. This translates to a photodiode responsivity of

80% is a typical quantum efficiency for InGaAs photodiodes at 1064 nm.

$$R_{\text{pd}} = \eta_{\text{pd}} \frac{q_e \, \lambda_{\text{laser}}}{h \, c} = 0.69 \, \frac{\text{A}}{\text{W}} \, , \tag{12}$$

where q_e is the electron charge, h is Planck's constant, and c is the speed of light.

The signal-to-noise ratio depends on the total equivalent input current noise of the amplifier, which consists of the input current noise ($\tilde{I}_{\text{pd}}$, set to 2 pA/$\sqrt{\text{Hz}}$) and intrinsic voltage noise of the amplifier ($\tilde{U}_{\text{pd}}$, set to 2 nV/$\sqrt{\text{Hz}}$) that is converted to current noise by the impedance of the photodiode. With an assumed photodiode capacitance $C_{\text{pd}} = 10 \, \text{pF}$ this impedance is given by

*A detailed calculation of the **voltage noise of the amplifier** considers the noise gain of the input stage that involves the value of the transimpedance resistor, which then cancels in the cited result.*

$$Z_{\text{pd}} = \frac{1}{2\pi \, C_{\text{pd}} \, f_{\text{het}}} = 884.22 \, \Omega \, . \tag{13}$$

The higher the heterodyne frequency f_{het} or capacitance, the lower the impedance becomes, which in turn will increase the resulting current noise of the transimpedance amplifier. The bandwidth of the photoreceiver is dictated by the heterodyne frequency. A more detailed discussion can be found in [71].

The various noise quantities in the signal add up differently depending on the number of photodiode segments used in the detection. In the following

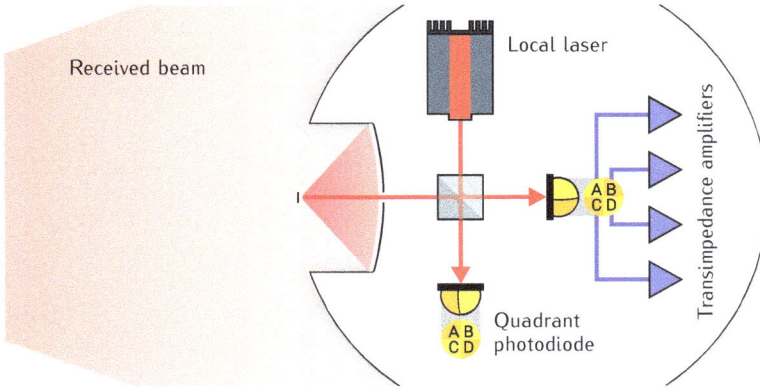

Figure 2.8: Two quadrant photodiodes (four segments each, one at each output port of a 50:50 beam splitter) are used to read out the heterodyne signal. Each segment is connected to a transimpedance amplifier.

I consider one pair of redundant quadrant photodiodes with four segments each ($N_{pd} = 4$) as illustrated in Figure 2.8. All equations are kept generally valid though and account for an arbitrary number of segments.

2.1.4 TEMPERATURE STABILITY

Some components will alter the overall optical path length or in general the phase of essential signals when a change in temperature occurs. While we assume a white path length noise over the measurement band for the telescope (see Section 2.2.3) we will use a more complex temperature noise model to calculate the influence on the optical bench as well as on some electronic and electro-optical components. Figure 2.9 shows a plot of the assumed temperature noise in Kelvin/$\sqrt{Hz}$ over Fourier frequency f. The web application allows to set a noise floor, two corner frequencies, and a lower and upper slope for each noise model.

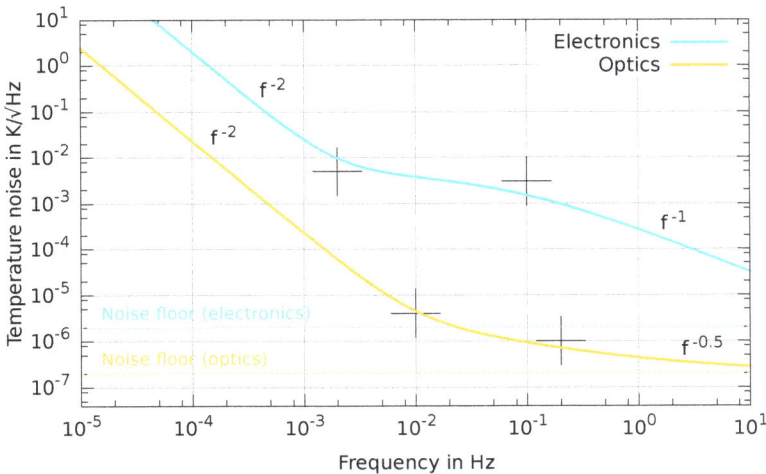

Figure 2.9: Temperature noise at the electronics and electro-optics (blue) and at the optical bench (yellow) in the significant heterodyne frequency range.

The blue trace in Figure 2.9 corresponds to the temperature noise, $\widetilde{T}_{el}\,(f)$, at the electronics and electro-optics which are usually distributed in boxes within the spacecraft, and results in noise levels of $5\,\text{mK}/\sqrt{\text{Hz}}$ and $3\,\text{mK}/\sqrt{\text{Hz}}$ at $f\,=\,2\times10^{-3}\,\text{Hz}$ and $f\,=\,1\times10^{-1}\,\text{Hz}$, respectively. The slopes below and above these corner frequencies are f^{-2} and f^{-1} with a constant noise floor of $2\times10^{-3}\,\text{mK}/\sqrt{\text{Hz}}$. The yellow trace corresponds to the temperature noise, $\widetilde{T}_{ob}\,(f)$, at the optical bench which is placed at the center of the spacecraft where the temperature is commonly more stable. We assume noise levels of $4\times10^{-3}\,\text{mK}/\sqrt{\text{Hz}}$ and $1\times10^{-3}\,\text{mK}/\sqrt{\text{Hz}}$ at $f\,=\,1\times10^{-2}\,\text{Hz}$ and $f=2\times10^{-1}\,\text{Hz}$ respectively. The slopes below and above these corner frequencies are f^{-2} and $f^{-0.5}$ with a constant noise floor of $2\times10^{-4}\,\text{mK}/\sqrt{\text{Hz}}$.

These values were chosen to keep temperature driven path length noise of the optical bench and the electronic and electro-optical components below shot noise and acceleration noise. Hence the specified temperature noise can be interpreted as temperature stability requirements.

2.2 DISPLACEMENT NOISE CONTRIBUTIONS

While we will read out phase shifts, $\delta\phi$, in the heterodyne signal, a more intuitive quantity is the apparent spacecraft displacement, δx, that corresponds to a measured phase shift. Since phase shifts in the individual laser beams are preserved in the heterodyne signal, the conversion $\delta\phi$ and δx is expressed by

$$\delta x = \frac{\lambda_{\text{laser}}}{2\pi} \times \delta\phi\,. \tag{14}$$

The same is true for the conversion between linear spectral densities of phase noise $\widetilde{\phi}$ (given in rad/$\sqrt{\text{Hz}}$) and displacement noise $\widetilde{x}$ (given in m/$\sqrt{\text{Hz}}$) which is used throughout this thesis.

There are multiple noise sources that are indistinguishable from an actual spacecraft displacement due to gravitational waves, any one of which potentially limits the observatory's sensitivity. In the following we will compute each displacement noise contribution individually.

2.2.1 READ-OUT NOISE

One important displacement noise contribution—and by design often the limiting one—is noise in the heterodyne signal read out, particularly noise in the electric current of the photo detector that measures the interference signal of received and local laser beams. The carrier-to-noise-density ratio C/N_0 (in units of power spectral density) can be used to calculate the resulting phase noise $\widetilde{\phi}_{r/o}$ in units of rad/$\sqrt{\text{Hz}}$ (linear spectral density):

$$\widetilde{\phi}_{r/o}\left[\frac{\text{rad}}{\sqrt{\text{Hz}}}\right] = \frac{1}{\sqrt{C/N_0}}\,. \tag{15}$$

In our case, C corresponds to the electrical signal power, and the amplitude $\sqrt{C}$ can be expressed as electric current

$$I_{\text{total}} = R_{\text{pd}} \frac{P_{\text{total}}}{2N_{\text{pd}}} , \qquad (16)$$

which is proportional to the time-dependent total incident optical power

$$P_{\text{total}} = \overbrace{P_{\text{local}} + P_{\text{rec}}}^{\text{DC term}} + \\ \underbrace{\overbrace{2\sqrt{\eta_{\text{het}} P_{\text{local}} P_{\text{rec}}}}^{\text{amplitude}} \overbrace{\sin\left(2\pi f_{\text{het}} t + \varphi\right)}^{\text{time dependence}}}_{\text{AC term (heterodyne beat note)}} \qquad (17)$$

where P_{local} is the power of the local laser. Dropping the DC term and the time dependence, the RMS electrical signal for the heterodyne beat note on one segment of a photodiode is found as

Variations in the laser power that affect this DC term are treated in Section 2.2.1.2.

$$I_{\text{signal, rms}} = \frac{1}{\sqrt{2}} R_{\text{pd}} \frac{2\sqrt{\eta_{\text{het}} P_{\text{local}} P_{\text{rec}}}}{2N_{\text{pd}}} . \qquad (18)$$

The factor $2N_{\text{pd}}$ accounts for the fact that there are two output ports of the 50:50 beam splitter that combines the received laser light with the local laser light, and each beam is distributed over N_{pd} segments of the photodiode.

This makes up half of Equation 15. The other half, N_0, corresponds to the noise-power spectral density. The single-sided linear spectral density $\sqrt{N_0}$ can be expressed as the electric current noise $\tilde{I}$ in units of $A/\sqrt{\text{Hz}}$. It is composed of

1. **shot noise**, the fluctuations of the number of photons detected,

2. **relative intensity noise**, the fluctuations in the laser power, and

3. **electrical noise**, the residual noise introduced by the transimpedance amplifier.

We will now determine the individual noise contribution for each component.

2.2.1.1 SHOT NOISE

For our purpose it is sufficient to compute the shot noise based on the DC term found in Equation 17 which leads to a total average DC photocurrent

$$I_{\text{dc}} \approx R_{\text{pd}} \frac{P_{\text{local}} + P_{\text{rec}}}{2N_{\text{pd}}} . \qquad (19)$$

With q_{e} as the electron charge the shot noise can now be expressed as

$$\tilde{I}_{\text{sn}} = \sqrt{2\, q_{\text{e}}\, I_{\text{dc}}} \approx \sqrt{2\, q_{\text{e}}\, R_{\text{pd}} \frac{P_{\text{local}} + P_{\text{rec}}}{2N_{\text{pd}}}} \qquad (20)$$

with minor corrections to be found in [72, 73]. Following Equation 15 the read-out noise due to shot noise is

$$\widetilde{\phi}_{r/o}^{sn} = \frac{\widetilde{I}_{sn}}{I_{signal}} = \sqrt{\frac{2N_{pd}\, q_e\, (P_{local} + P_{rec})}{R_{pd}\, \eta_{het}\, P_{local}\, P_{rec}}} \tag{21}$$

for each photodiode segment. For sufficient high values of P_{local}/P_{rec}, a change in total local laser power will have no influence on the shot noise in the signal read-out.

Shot noise is a non-correlated contribution between different photodiodes and their segments, hence averaging over all N_{pd} segments will improve the signal quality by a factor of $\sqrt{N_{pd}}$ so it becomes independent of the number of segments (single-element vs. quadrant photodiode):

$$\langle \widetilde{\phi}_{r/o}^{sn} \rangle = \frac{1}{\sqrt{N_{pd}}}\, \widetilde{\phi}_{r/o}^{sn} = \sqrt{\frac{2\, q_e\, (P_{local} + P_{rec})}{R_{pd}\, \eta_{het}\, P_{local}\, P_{rec}}}\;. \tag{22}$$

2.2.1.2 RELATIVE INTENSITY NOISE

The relative intensity noise, RIN, as described in Section 2.1.3 is assumed of equal magnitude but uncorrelated between both laser beams. It couples directly to the photocurrent and adds quadratically:

$$\begin{aligned}
\widetilde{I}_{rin} &= \sqrt{\left(R_{pd}\frac{P_{local}}{2N_{pd}}RIN\right)^2 + \left(R_{pd}\frac{P_{rec}}{2N_{pd}}RIN\right)^2} \\
&= R_{pd}\frac{\sqrt{P_{local}^2 + P_{rec}^2}}{2N_{pd}}\, RIN\;.
\end{aligned} \tag{23}$$

Consequently, the read-out noise due to relative intensity noise is

$$\widetilde{\phi}_{r/o}^{rin} = \frac{\widetilde{I}_{rin}}{I_{signal}} = RIN\, \sqrt{\frac{P_{local}^2 + P_{rec}^2}{2\eta_{het}\, P_{local}\, P_{rec}}} \tag{24}$$

for each photodiode segment. It is generally independent of the number of segments and the photodiode responsivity. In contrast to shot noise, a higher local laser power will <u>increase</u> the influence of relative intensity noise in the signal read-out.

Since the relative intensity noise is correlated in both beam splitter outputs and on each photodiode segment, averaging over photodiodes or N_{pd} segments does not yield any improvements in the signal quality:

$$\langle \widetilde{\phi}_{r/o}^{rin} \rangle = \widetilde{\phi}_{r/o}^{rin} = RIN\, \sqrt{\frac{P_{local}^2 + P_{rec}^2}{2\eta_{het}\, P_{local}\, P_{rec}}}\;. \tag{25}$$

2.2.1.3 ELECTRICAL NOISE

The photodiode preamplifier (transimpedance amplifier) shows input current noise, $\widetilde{I}_{pd}$, as well as uncorrelated voltage noise, $\widetilde{U}_{pd}$, that can be converted to equivalent input current noise $\widetilde{I}_{tia} = \widetilde{U}_{pd}/Z_{pd}$ using the photodiode's impedance Z_{pd}. Both contributions add quadratically.

$$\widetilde{I}_{el} = \sqrt{\widetilde{I}_{pd}^2 + \widetilde{I}_{tia}^2} = \sqrt{\widetilde{I}_{pd}^2 + \left(\frac{\widetilde{U}_{pd}}{Z_{pd}}\right)^2} \tag{26}$$

The read-out noise due to electronic noise is then given by

$$\widetilde{\phi}_{r/o}^{el} = \frac{\widetilde{I}_{el}}{I_{signal}} = N_{pd} \frac{\sqrt{2}}{R_{pd}} \sqrt{\frac{\widetilde{I}_{pd}^2 + \left(\frac{\widetilde{U}_{pd}}{Z_{pd}}\right)^2}{\eta_{het} P_{local} P_{rec}}} \tag{27}$$

for each photodiode segment. Here, a higher local laser power will <u>reduce</u> the influence of electronic noise in the signal read-out.

Electronic noise is a non-correlated contribution between different photodiodes and their segments, hence averaging over all N_{pd} segments will improve the signal quality by a factor of $\sqrt{N_{pd}}$. As a result, the influence of electronic noise in the signal read-out scales by $\sqrt{N_{pd}}$ since each channel is amplified individually:

$$\left\langle \widetilde{\phi}_{r/o}^{el} \right\rangle = \frac{1}{\sqrt{N_{pd}}} \widetilde{\phi}_{r/o}^{el} = \frac{\sqrt{2 N_{pd}}}{R_{pd}} \sqrt{\frac{\widetilde{I}_{pd}^2 + \left(\frac{\widetilde{U}_{pd}}{Z_{pd}}\right)^2}{\eta_{het} P_{local} P_{rec}}}. \tag{28}$$

2.2.1.4 OPTIMAL LOCAL LASER POWER

As mentioned above, the influence of the different read-out noise contributions scales differently with local laser power P_{local}. Figure 2.10 shows the total read-out noise

$$\left\langle \widetilde{\phi}_{r/o}^{total} \right\rangle = \sqrt{\left\langle \widetilde{\phi}_{r/o}^{sn} \right\rangle^2 + \left\langle \widetilde{\phi}_{r/o}^{rin} \right\rangle^2 + \left\langle \widetilde{\phi}_{r/o}^{el} \right\rangle^2} \tag{29}$$

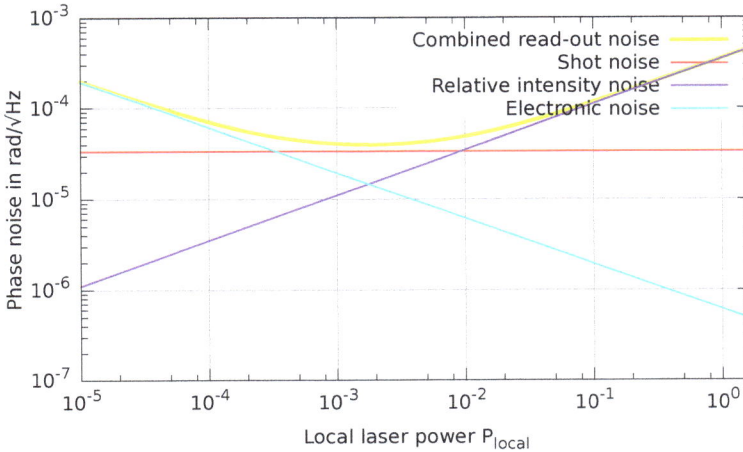

Figure 2.10: Linear spectral density of combined read-out phase noise (green) and its individual contributions over local laser power P_{local}.

as well as the individual contributions for the given parameters plotted over local laser power. A minimum of this function can be found for $P_{local} = 1.75 \times 10^{-3}$ Watts.

Before we can compute the absolute values for the different read-out noise contributions, we have to consider that in reality the laser beams are phase modulated and carry additional information in sidebands. As a result, the heterodyne signal now consists of a carrier beat note and multiple sideband beat notes. These sidebands consume some of the total signal power. In the present case we require each of the two first-order sidebands to hold 7.5% of the carrier's power [74, p. 28]. Additional signal modulation used for inter-spacecraft data transfer and ranging is assumed to contain approximately 1% of the signal power as shown in [75]. This very small additional optical power drain is currently not considered by the web application and thus can be ignored at this point. The resulting frequency spectrum can be calculated using Bessel functions of the first kind (J_0, J_1, J_2, ...).

Figure 2.11 shows the power for the carrier ($J_0(m)^2$) and the first- and second-order sidebands ($J_1(m)^2$, $J_2(m)^2$) as fractions of the total power as a function of the modulation depth m. The desired ratio between carrier and first-order sideband of 7.5% occurs at $m = 0.53$ rad. Accordingly the RMS electrical signal for the carrier beat note has to be written as

$$
\begin{aligned}
I_{carrier} &= \frac{1}{\sqrt{2}} R_{pd} \frac{2\sqrt{\eta_{het} J_0(m)^2 P_{local} J_0(m)^2 P_{rec}}}{2N_{pd}} \\
&= J_0(m)^2 I_{signal} \, .
\end{aligned}
\tag{30}
$$

We must apply this reduced carrier signal level to the read-out noise calculations. As derived from Equations 21, 24, and 27, the individual noise contri-

High-power first-order sidebands that result in a modulation depth $m > 1$ will additionally be accompanied by higher order sidebands. This should be avoided since these sidebands are not used but nevertheless reduce the overall signal power.

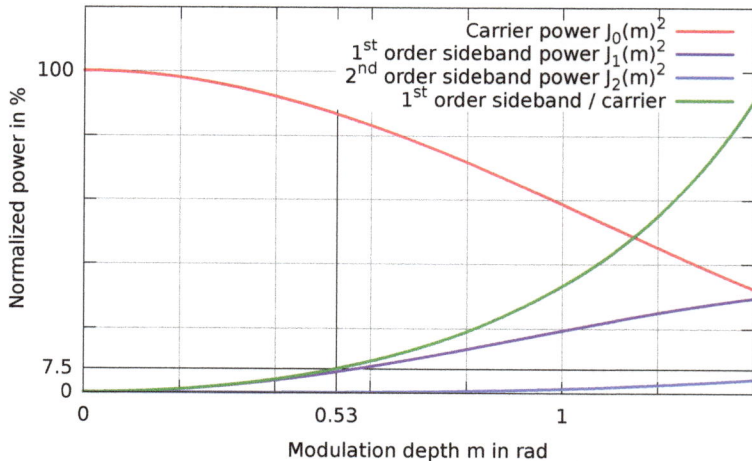

Figure 2.11: Carrier and first- and second-order sidebands (normalized power over modulation depth m). The desired ratio between carrier and first-order sideband (green trace) of 7.5% occurs at $m = 0.53$ rad as indicated.

butions are simply increased by the factor $1/J_0(m)^2 = 1.15$. Converted to displacement noise (see Equation 14) we finally obtain

$$
\begin{aligned}
\left\langle \tilde{x}_{\text{r/o}}^{\text{sn}} \right\rangle_{\text{carrier}} &= \frac{\lambda_{\text{laser}}}{2\pi} \frac{1}{J_0(m)^2} \left\langle \tilde{\phi}_{\text{r/o}}^{\text{sn}} \right\rangle \\
&= 6.58 \times 10^{-12} \frac{\text{m}}{\sqrt{\text{Hz}}} \,,
\end{aligned}
\tag{31}
$$

$$
\begin{aligned}
\left\langle \tilde{x}_{\text{r/o}}^{\text{rin}} \right\rangle_{\text{carrier}} &= \frac{\lambda_{\text{laser}}}{2\pi} \frac{1}{J_0(m)^2} \left\langle \tilde{\phi}_{\text{r/o}}^{\text{rin}} \right\rangle \\
&= 2.85 \times 10^{-12} \frac{\text{m}}{\sqrt{\text{Hz}}} \,, \text{ and}
\end{aligned}
\tag{32}
$$

$$
\begin{aligned}
\left\langle \tilde{x}_{\text{r/o}}^{\text{el}} \right\rangle_{\text{carrier}} &= \frac{\lambda_{\text{laser}}}{2\pi} \frac{1}{J_0(m)^2} \left\langle \tilde{\phi}_{\text{r/o}}^{\text{el}} \right\rangle \\
&= 2.86 \times 10^{-12} \frac{\text{m}}{\sqrt{\text{Hz}}} \,.
\end{aligned}
\tag{33}
$$

From the values above we conclude that the total read-out noise in the carrier signal,

$$
\begin{aligned}
\left\langle \tilde{x}_{\text{r/o}}^{\text{total}} \right\rangle_{\text{carrier}} &= \frac{\lambda_{\text{laser}}}{2\pi} \frac{1}{J_0(m)^2} \left\langle \tilde{\phi}_{\text{r/o}}^{\text{total}} \right\rangle \\
&= 7.73 \times 10^{-12} \frac{\text{m}}{\sqrt{\text{Hz}}} \,,
\end{aligned}
\tag{34}
$$

is limited by shot noise as desired for a carefully designed gravitational wave observatory. This can also be seen in Figure 2.10. The value of $\left\langle \tilde{x}_{\text{r/o}}^{\text{total}} \right\rangle_{\text{carrier}}$ is equivalent to a phase noise of 4.56×10^{-5} rad$/\sqrt{\text{Hz}}$. One usually aims to keep additional phase fluctuations of the signal as well as all noise introduced during phase measurement, post-processing, and data analysis well below this level.

2.2.2 CLOCK NOISE

To measure the phase of the carrier signal, the analog output from the trans-impedance amplifier is digitized by an analog-to-digital converter (ADC) that is triggered by a reference oscillator (system clock). Here, timing noise $\tilde{t}$ leads to phase noise

$$
\tilde{\phi} = 2\pi f \tilde{t}
\tag{35}
$$

in the digital representation of the signal. For the measurement of a signal with frequency $f = f_{\text{het}}$ a timing stability of $\tilde{t} < 4.03 \times 10^{-13}$ s$/\sqrt{\text{Hz}}$ would be required to stay below the above calculated total carrier signal read-out phase noise. Unfortunately, ADCs and oscillators that stable do not exist. To deal with the excess noise, additional signals called 'pilot tones' are introduced. Within one spacecraft, a common pilot tone (at frequency f_p outside

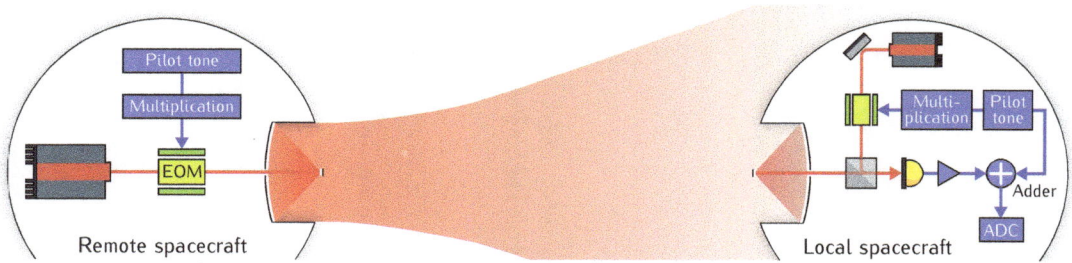

Figure 2.12: Pilot tone distribution for a single link of the observatory. At the remote spacecraft, the pilot tone frequency is multiplied and the signal is modulated onto the outgoing laser beam by an electro-optic phase modulator (EOM). A separate pilot tone on the local spacecraft is modulated onto the local laser to compare the two pilot tones in the sideband beat note of the heterodyne signal. To suppress the influence of ADC timing jitter, the local pilot tone is added to the heterodyne signal and used as a reference.

the heterodyne signal bandwidth) is combined with the analog output from each transimpedance amplifier. Both signals are digitized simultaneously by the same ADC in each channel. Thus we can use the pilot tone as a reference to suppress the influence of ADC timing jitter on the digitized carrier signal.

In addition, the pilot tones of different spacecraft are modulated on the outgoing laser beams by electro-optic phase modulators (EOMs) as illustrated in Figure 2.12. The affiliated first-order sidebands (as already mentioned in Section 2.2.1.4) each hold 7.5% of the carrier's power and result in sideband beat notes in the heterodyne signal. These additional beat notes (which must fall within the heterodyne signal bandwidth) are correlated with the differential phase noise between the corresponding remote and local pilot tones. Thus we can compare the pilot tones between all spacecraft and construct a constellation wide common reference during post-processing. As a result, a specific timing stability of the individual system clocks is no longer required, but we now may be limited by

1. **read-out noise** in the sideband beat notes, and

2. excess phase noise introduced by components in the **pilot tone transmission chain**.

I will now calculate the corresponding displacement noise contributions for both influences.

2.2.2.1 SIDEBAND SIGNAL READ–OUT NOISE

Since the RMS electrical signal for the sideband beat note

$$I_{\text{sideband}} = 7.5\% \times I_{\text{carrier}} = J_1(m)^2 \, I_{\text{signal}} \tag{36}$$

The **timing jitter** conservation of frequency multipliers (Section 5.2.2) and dividers (Section 4.3.2) stands in contrast to the mixing process in, e.g., heterodyne interferometry or electronic mixers, which maintains phase information (see Section 2.1.2).

is smaller than the carrier signal (compare Equation 30), the read-out phase noise for the sideband signal will be much higher (compare Equation 34). To reduce the impact of read-out noise on the sideband signals, we boost the desired signal—which is the pilot tone's phase information—before modulating it onto the laser beam. This can be done by frequency multipliers as they conserve timing jitter and hence lead to an amplification of phase jitter by the frequency multiplication (or signal amplification) factor f_{mod}/f_p, where f_{mod} represents the actual modulation frequency [76].

Consequently, the total read-out noise for one first-order sideband beat note expressed in phase noise,

$$\left\langle \widetilde{\phi}_{\text{r/o}}^{\text{total}} \right\rangle_{\text{sideband}} = \frac{f_{\text{p}}}{f_{\text{mod}}} \frac{1}{J_1(m)^2} \left\langle \widetilde{\phi}_{\text{r/o}}^{\text{total}} \right\rangle , \tag{37}$$

scales with the inverse of the signal amplification factor. Furthermore, the equivalent displacement noise, according to Equation 14, scales with the ratio of the maximum heterodyne frequency f_{het} to the pilot tone frequency f_p as

$$\left\langle \widetilde{x}_{\text{r/o}}^{\text{total}} \right\rangle_{\text{sideband}} = \frac{\lambda_{\text{laser}}}{2\pi} \frac{f_{\text{het}}}{f_p} \left\langle \widetilde{\phi}_{\text{r/o}}^{\text{total}} \right\rangle_{\text{sideband}} \tag{38}$$

since all measurements at frequency f_{het} are referenced to the pilot tone. The higher the pilot tone frequency, the less phase jitter of the pilot tone impacts the measurement of a signal. The higher the signal frequency, the more it is influenced by phase jitter of the pilot tone.

In conclusion, summing up the total read-out noise for both sidebands quadratically (factor $1/\sqrt{2}$), we get

$$\begin{aligned}
\left\langle \widetilde{x}_{\text{r/o}}^{\text{total}} \right\rangle_{\text{sidebands}} &= \frac{1}{\sqrt{2}} \frac{\lambda_{\text{laser}}}{2\pi} \frac{f_{\text{het}}}{f_{\text{mod}}} \frac{1}{J_1(m)^2} \left\langle \widetilde{\phi}_{\text{r/o}}^{\text{total}} \right\rangle \\
&= 5.46 \times 10^{-13} \, \frac{\text{m}}{\sqrt{\text{Hz}}}
\end{aligned} \tag{39}$$

for a modulation frequency of $f_{\text{mod}} = 2.40\,\text{GHz}$. This value represents the excess noise of the observatory introduced by the imperfect synchronization of the pilot tones between the different spacecraft due to the noisy read-out of the sideband signal. The pilot tone frequency f_p does not influence this noise level but might be of importance during the actual phase measurement and for generating the modulation signal.

2.2.2.2 PILOT TONE TRANSMISSION CHAIN NOISE

Another source of excess noise results from an imperfect pilot tone fidelity, that is when the phase of the modulation sidebands differs from the phase of the corresponding pilot tones used for the ADC timing jitter correction. Components in the pilot tone transmission chain might shift the phase of the pilot tone (in the electrical signal) or of the sidebands (in the optical signal). There are many components involved that can potentially limit the observatory's sensitivity in this way. One of this thesis' main accomplishments is a thorough study of these components and the development of a compatible pilot tone generation and distribution chain. Noise figures for individual components used in this section are based on this detailed investigation presented in Chapter 5.

As illustrated in Figure 2.13 (blue items) the electrical transmission chain contains a number of power splitters, the electronic addition stages for the

Figure 2.13: The pilot tone must be phase stable to the sidebands that are modulated onto the outgoing laser beam by an EOM. It is combined with the heterodyne signals and used as a reference to suppress timing jitter. Components in the transmission line from the pilot tone generation to the ADCs (like power splitters and adders) and to the transmitting telescope (like fiber amplifiers (FA) and optical fibers) might add phase noise.

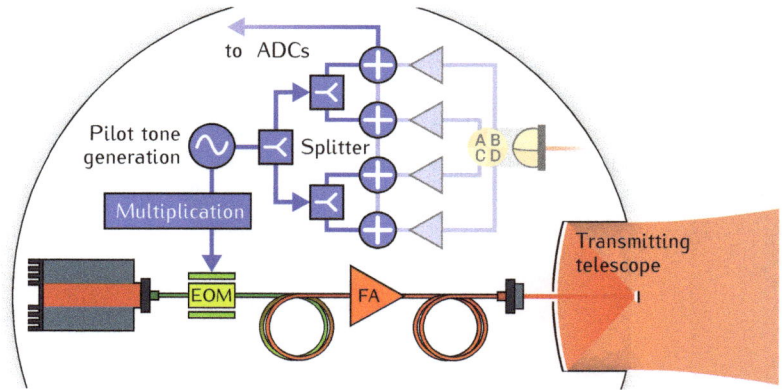

pilot tone and the heterodyne signal, as well as the frequency multiplier or divider, possibly in multiple instances. Since phase noise introduced by any of these components depends on the actual pilot tone frequency, the combined noise introduced by all electrical components in the pilot tone transmission chain, $\widetilde{t}_{el} = 4 \times 10^{-14}\,\text{s}/\sqrt{\text{Hz}}$, is given in frequency independent units of timing jitter [74]. This translates to a equivalent displacement noise of up to

$$\widetilde{x}_{tml}^{el} = \lambda_{laser}\, f_{het}\, \widetilde{t}_{el} = 7.66 \times 10^{-13}\, \frac{\text{m}}{\sqrt{\text{Hz}}} \tag{40}$$

when the heterodyne frequency reaches its maximum value.

Keep in mind that the above value, like most noise figures given in this chapter, depend on the temperature stability. Actual dependencies for individual components may change with temperature and can also partly cancel each other. Thus a complete timing noise model for all electrical components would turn out to be quite complex.

Also electrical cables connecting the different components shift the phase of the pilot tone and modulation signal in accordance with temperature due to a number of effects, among others a change in the dielectric constant of the inner insulator and a change in the cables' dimension. This will alter the velocity of propagation and the electrical length of the transmission line respectively.

The absolute phase shift depends on the actual frequency of the signal passed along the cable, and different frequencies (f_p, f_{mod}) are involved. However, with a thermal stability of the electrical cables given per meter and gigahertz, we can calculate an equivalent displacement noise level independent of the signal frequency. This thermal stability is assumed to be

$$\left(\frac{\delta\phi}{\delta T}\right)_{cables} = 7\, \frac{\text{mrad}}{\text{K}}\, \frac{1}{\text{m} \times \text{GHz}}\,. \tag{41}$$

It leads to a noise due to temperature shifts in the electrical cables [76] that is given by the temperature noise at the electronics and electro-optics $\widetilde{T}_{el}(f)$ (compare Section 2.1.4) of

$$
\begin{aligned}
\widetilde{x}_{tml}^{cables}(f) &= \frac{\lambda_{laser}}{2\pi} f_{het} \widetilde{T}_{el}(f) \, l_{cables} \left(\frac{\delta\phi}{\delta T}\right)_{cables} \\
&= 4.27 \times 10^{-11} \frac{m}{K} \times \widetilde{T}_{el}(f) \; .
\end{aligned}
\tag{42}
$$

That noise changes with Fourier frequency f. The length of unmatched electrical cables was assumed to be $l_{cables} = 2\,m$. This is a rather arbitrary number but shows how even a short electrical transmission line between the EOM and the ADCs can result in a significant amount of displacement noise.

In Equation 42 the signal frequency (f_p, f_{mod}) canceled with parts of the corresponding scaling factor introduced by Equation 38, and only the maximum heterodyne frequency f_{het} remains.

Likewise, the influence of optical fibers that pass the modulated laser light from the EOM to the transmitting telescope (see Figure 2.13) can be calculated. Here, the modulation signal is phase shifted with temperature due to a change in the fibers' dimension and refractive index. For a thermal stability of the fibers given per meter and gigahertz [76],

$$
\left(\frac{\delta\phi}{\delta T}\right)_{fibers} = 1 \, \frac{mrad}{K} \, \frac{1}{m \times GHz} \; ,
\tag{43}
$$

and a total fiber length of $l_{fibers} = 5\,m$, the equivalent displacement noise due to temperature shifts in the optical fibers is

$$
\begin{aligned}
\widetilde{x}_{tml}^{fibers}(f) &= \frac{\lambda_{laser}}{2\pi} f_{het} \widetilde{T}_{el}(f) \, l_{fibers} \left(\frac{\delta\phi}{\delta T}\right)_{fibers} \\
&= 1.52 \times 10^{-11} \frac{m}{K} \times \widetilde{T}_{el}(f) \; .
\end{aligned}
\tag{44}
$$

This noise is given by the same temperature noise at the electronics and electro-optics $\widetilde{T}_{el}(f)$, compare Section 2.1.4.

Finally, the two electro-optic components that sit in the optical transmission chain, namely the EOM and a fiber amplifier (FA) that boosts the laser power to $> P_{tel}$ before passing it to the telescope, can influence the phase of the sidebands. The performance of both devices depends on the absolute temperature, light power, temperature stability and other environmental influences and should be subject to a separate study. We assumed a phase noise of $\widetilde{\phi}_{eom} = 3 \times 10^{-4}\,rad/\sqrt{Hz}$ for the EOM and $\widetilde{\phi}_{fa} = 6 \times 10^{-4}\,rad/\sqrt{Hz}$ for the FA, valid at the modulation frequency f_{mod}.

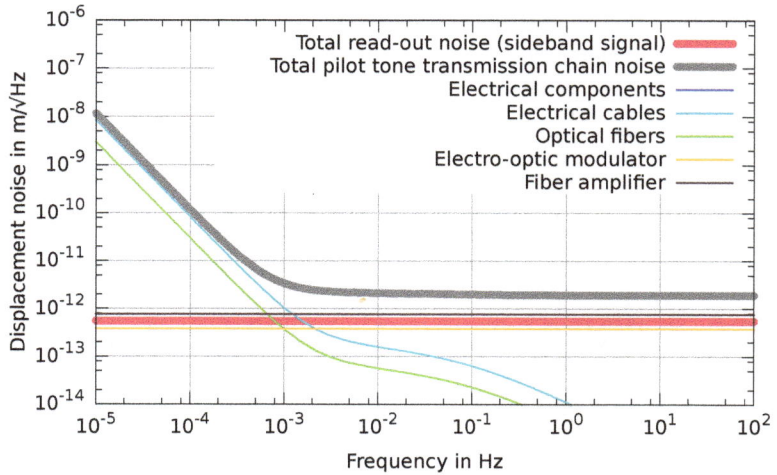

Figure 2.14: Pilot tone transmission chain noise contributions plotted over Fourier frequency, including the sideband read-out noise (for both sideband signals combined) and the individual pilot tone transmission chain noise components.

Since all noise sources are temperature dependent, we conservatively add the individual figures linearly and come up with a total pilot tone transmission chain noise of

$$\tilde{x}_{\text{tml}}^{\text{total}}(f) = \tilde{x}_{\text{tml}}^{\text{el}} + \tilde{x}_{\text{tml}}^{\text{cables}}(f) + \tilde{x}_{\text{tml}}^{\text{fibers}}(f)$$
$$+ \frac{\lambda_{\text{laser}}}{2\pi}\left(\tilde{\phi}_{\text{eom}} + \tilde{\phi}_{\text{fa}}\right). \tag{45}$$

Figure 2.14 shows all noise contributions individually over Fourier frequency f. At low frequencies, the importance of the consideration of temperature noise becomes obvious since it clearly dominates the equivalent displacement noise.

2.2.3 OPTICAL PATH LENGTH NOISE

The optical telescopes are naturally within the optical path of the interferometer and jitter of the optical path length through the telescope is of direct concern. The dimensional jitter is caused by temperature noise at the telescope, but it is hard to model because of a strong temperature gradient. While the primary mirror usually lies deep within the spacecraft and could be close to room temperature, the secondary mirror is more exposed to outer space and may be as cold as a few Kelvin. Dimensional stability investigations for carbon fiber reinforced polymer and ultra-low expansion glass-ceramic structures predict a path length noise smaller than $\tilde{x}_{\text{opn}}^{\text{tel}} = 1\,\text{pm}/\sqrt{\text{Hz}}$ down to frequencies of 1 mHz [77].

To allow for easier component testing on ground, spacecraft are usually designed to provide an internal temperature close to room temperature.

Additionally, a change in temperature of the optical bench – which consists of fused silica components bonded to a base plate made thermally-compensated glass-ceramic – results in a uniform expansion of the material that leads to a phase shift in the heterodyne signals. If more than one interferometer is located on a single optical bench, this effect will only cancel out if the path lengths on the optical bench are matched for all interferometers. Otherwise, the phase noise due to temperature fluctuations will not cancel completely. Instead, there will be a coupling factor that scales with the difference in the optical path lengths of at least two interferometers involved.

As discussed in Section 2.1.3, a dedicated inter-spacecraft interferometer is needed to utilize the full power of the received beam and minimize the influence of read-out noise. In this read-out scheme, the influence of the optical path length difference in the combination of any two inter-spacecraft interferometers for one full observatory arm cancels each other. Accordingly the relevant path length difference is the one between the two additional interferometers required to determine the proof mass displacement: the proof mass interferometer and the reference interferometer, compare Figure 2.7.

One must distinguish between the path length difference within fused silica, $\mathrm{OPD_{fs}}$, and the path length difference on the optical bench itself, $\mathrm{OPD_{ob}}$, where the beam translates in vacuum. We assume values of $\mathrm{OPD_{fs}} = 29\,\mathrm{mm}$ and $\mathrm{OPD_{ob}} = 565\,\mathrm{mm}$. The latter is the total path length difference on the optical bench including light paths within fused silica optics, so that the significant path length difference on the glass-ceramic base plate is given by $\mathrm{OPD_{ob}} - \mathrm{OPD_{fs}}$. We can now calculate the equivalent displacement noise contributions due to the path length imbalances. With the given temperature noise at the optical bench, $\widetilde{T}_{\mathrm{ob}}\,(f)$, and the coefficient of thermal expansion of glass-ceramic, $\alpha_{\mathrm{ule}} = 2 \times 10^{-8}\,\mathrm{m/K}$, the path length noise of the base plate can be expressed as

$$\widetilde{x}^{\mathrm{ule}}_{\mathrm{opn}}\,(f) = \widetilde{T}_{\mathrm{ob}}\,(f) \times (\mathrm{OPD_{ob}} - \mathrm{OPD_{fs}}) \times \alpha_{\mathrm{ule}}\,. \tag{46}$$

The description of the path length noise introduced by the fused silica components is more complex since the laser beam is passing through those components and not through vacuum. Thus we have to consider the refractive index of fused silica, $n_{\mathrm{fs}} = 1.45$, as well as its change with temperature, $dn_{\mathrm{fs}}/dT = 1.10 \times 10^{-6}\,/\mathrm{K}$. With the coefficient of thermal expansion of fused silica, $\alpha_{\mathrm{fs}} = 5.50 \times 10^{-7}\,\mathrm{m/K}$, the equivalent displacement noise can then be expressed as

The given values are only valid for a wavelength of 1064 nm at room temperature.

$$\widetilde{x}^{\mathrm{fs}}_{\mathrm{opn}}\,(f) = \widetilde{T}_{\mathrm{ob}}\,(f) \times \mathrm{OPD_{fs}} \left(\alpha_{\mathrm{fs}}\,(n_{\mathrm{fs}} - 1) + \frac{dn_{\mathrm{fs}}}{dT} \right)\,. \tag{47}$$

In the above equation we use the difference of the refractive index of fused silica and vacuum, $n_{\mathrm{fs}} - 1$. This is due to the fact that an increase in the path length for light passing through fused silica simultaneously decreases the path length in vacuum.

While both path length noise contributions of the optical bench add linearly— since they are the result of the very same temperature fluctuations—the optical path length noise of the telescope relates to an uncorrelated temperature

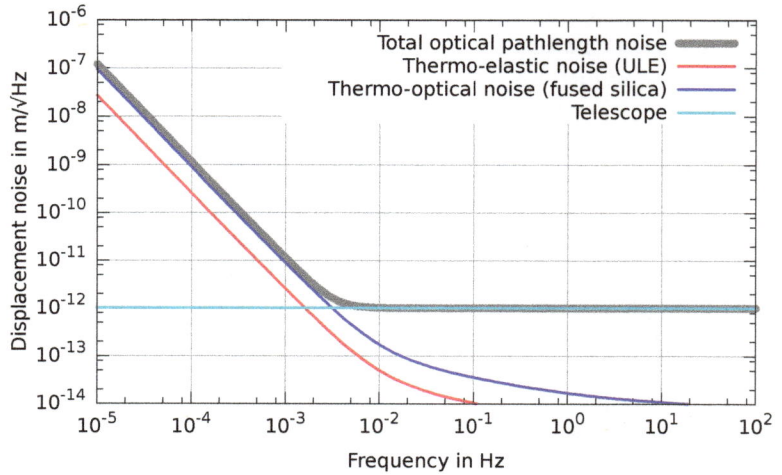

Figure 2.15: Total optical path length noise and individual contributions from optical bench and telescope. The summation of the contributions is described in Equation 48.

noise and hence adds quadratically. Thus the total optical path length noise has to be written as

$$\tilde{x}_{\text{opn}}^{\text{total}}(f) = \sqrt{\underbrace{\left[\tilde{x}_{\text{opn}}^{\text{ule}}(f) + \tilde{x}_{\text{opn}}^{\text{fs}}(f)\right]^2}_{\text{optical bench}} + \underbrace{\left(\tilde{x}_{\text{opn}}^{\text{tel}}\right)^2}_{\text{telescope}}}. \tag{48}$$

All optical path length noise contributions and the total optical path length noise are plotted as a function of the Fourier frequency f in Figure 2.15.

Other sources of optical path length noise, such as a non-uniform change in temperature, temperature gradients, and tilt-to-path length coupling [78], are neglected by the web application. These contributions are either specific to the detailed mission design and hence hard to generalize, or based on complex coupling mechanisms and hence difficult to predict. In a detailed mission design one needs to evaluate any of these influences in detail.

2.2.4 ACCELERATION NOISE

Residual forces on the proof masses, like Coulomb forces induced from imperfect cancellation of charges, surface effects, and residual gas pressure, result in a spurious acceleration of the proof masses [79]. Generally, a white acceleration noise of $3 \times 10^{-15}\,\text{m s}^{-2}/\sqrt{\text{Hz}}$ is assumed. This can be described as frequency dependent displacement noise by

$$\tilde{x}_{\text{acc}}(f) = 3 \times 10^{-15}\,\frac{\text{m/s}^2}{\sqrt{\text{Hz}}} \times \frac{1}{(2\pi f)^2}. \tag{49}$$

In reality this function might be more complex due to the vast number of different effects acting on the proof masses [80–83]. The above value is just a rough estimate and does not include, for example, a shift in the local gravitational field due to spacecraft position jitter. Gravitational reference sensors

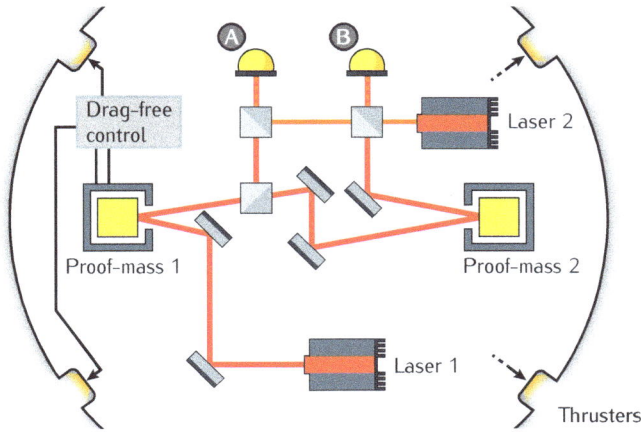

Figure 2.16: Lasers are used to read out the motion of the first proof mass (detector A) and the differential motion between both masses (detector B). While one proof mass is drag-free and the spacecraft follows its position by micro-thrusters, the other one is free floating except for a very gentle electrostatic suspension to prevent runaway.

that encapsulate the proof masses and are designed to keep this noise contribution to a minimum cannot be tested to full extent on ground [84].

On this account, LISA Pathfinder, an ESA space mission scheduled to launch in 2015 at the time of writing, will test the gravitational reference sensors under realistic conditions [85, 86]. Featuring one observatory arm shrunk to 40 cm in length and two proof masses all embedded within one spacecraft, this mission will be able to test all disturbances that act locally inside the spacecraft. The spacecraft will follow one freely floating proof mass by reading out its position with capacitive sensors and laser interferometry as shown in Figure 2.16. The spacecraft's position is actuated with micro-Newton thrusters. The second proof mass is kept in position relative to the spacecraft by capacitive actuators. In that sense LISA Pathfinder is a highly accurate gravimetric sensor that measures differences in the gravitational force between the two proof masses, i.e., a gravity gradiometer. Figure 2.17 shows the fully assembled LISA Pathfinder spacecraft with the propulsion

The spacecraft is now drag-free since its position is **actuated** to compensate outside influences. The one freely floating proof-mass acts as reference and ideally is only influenced by gravity.

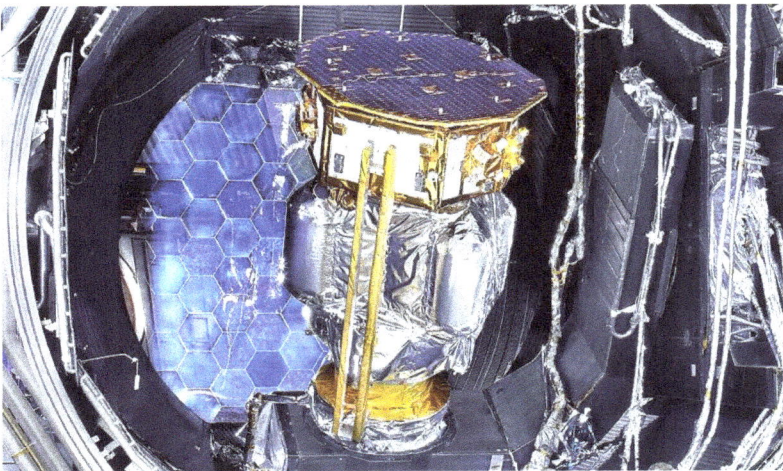

Figure 2.17: The LISA Pathfinder spacecraft with attached propulsion module inside a space condition simulation chamber at IABG (Germany) in August 2011. credit: © ESA/Airbus/IABG

module attached during environmental and performance tests inside a space simulator in August 2011. It will perform its experimental schedule at the L1 Lagrangian point for almost three months. Results will be available immediately, through real-time data analysis.

Acceleration noise is potentially limiting the observatory's sensitivity. Thus the LISA Pathfinder mission is considered to be one of the most important milestones towards a gravitational wave observatory in space. It will verify current acceleration noise models [87] and potentially result in improved and more realistic models.

2.2.5 METROLOGY AND DATA PROCESSING

The individual inter-spacecraft interferometers place one arm inside the spacecraft while the arm sensitive to gravitational waves is spread across two spacecraft, which results in a huge arm length difference that is equal to the spacecraft separation distance L_{arm}. As in any unequal-arm Michelson interferometer, the laser frequency noise $\widetilde{\nu}_{\text{pre}}$ of the pre-stabilized laser at frequency $\nu = c/\lambda_{\text{laser}}$ directly translates to displacement noise with

$$\widetilde{x}_{\text{ms}}^{\text{lfn}} = L_{\text{arm}} \times \frac{\widetilde{\nu}_{\text{pre}}}{\nu} = 2.06 \times 10^{-3} \, \frac{\text{m}}{\sqrt{\text{Hz}}} \, . \tag{50}$$

In the construction of the virtual Michelson interferometer where multiple individual interferometers are combined, only the difference in the arm length between the different spacecraft (roughly 1% of the total arm length) is of concern. Yet even this reduced noise level would dominate the entire observatory.

Such noise can be suppressed by a data post-processing technique called time-delay interferometry (TDI) [88, 89]. Here, signals from different interferometers are time-shifted and combined in such a way that laser frequency noise cancels to the greatest extent. This only works if A) we read out all beat notes in the heterodyne signal with sufficient precision, B) we have accurate knowledge of the inter-spacecraft separation distance, and C) we have precise time stamps of all measurements with respect to a constellation wide clock. The latter information will be used to determine the correct time-shifts in post-processing. It is gained by a combination of

1. **spacecraft position triangulation** by the Deep Space Network,

2. **ranging** with delayed pseudo random noise (PRN) codes modulated onto the laser beams [75, 90], and

3. **raw data pre-processing** by Kalman filters to recover the ranging information and base all measurements on a common reference frequency [91].

Everything considered, we assume that the knowledge of the absolute spacecraft separation is better than within $L_{\text{ranging}} = 0.10\,\text{m}$. We can thus calculate the equivalent displacement noise by simply adapting Equation 50 and get

$$\widetilde{x}_{\text{ms}}^{\text{tdi}} = L_{\text{ranging}} \times \frac{\widetilde{\nu}_{\text{pre}}}{\nu} = 1.03 \times 10^{-13}\,\frac{\text{m}}{\sqrt{\text{Hz}}}\,. \tag{51}$$

On top of that we assume an ancillary phase error in the signal read-out of $6\,\mu\text{rad}/\sqrt{\text{Hz}}$ at the maximum heterodyne frequency that is caused by the phasemeter (see Section 5.4.3 and [74, p. 23]). This translates to a displacement noise equivalent of

$$\widetilde{x}_{\text{ms}}^{\text{pm}} = 1.02 \times 10^{-12}\,\frac{\text{m}}{\sqrt{\text{Hz}}}\,. \tag{52}$$

While this read-out noise shows up in every single data stream, the ranging accuracy only comes into play when multiple links are combined. Technically speaking, each individual link is still limited by the noise level calculated in Equation 50. Nevertheless, for reasons of simplification, we add a metrology system and data processing noise level of

$$\widetilde{x}_{\text{ms}}^{\text{total}} = \sqrt{(\widetilde{x}_{\text{ms}}^{\text{pm}})^2 + (\widetilde{x}_{\text{ms}}^{\text{tdi}})^2} = 1.02 \times 10^{-12}\,\frac{\text{m}}{\sqrt{\text{Hz}}} \tag{53}$$

to the total displacement noise of each link. By doing so, we can compare all displacement noise contributions, summarized in Figure 2.18, and determine the limiting influences.

The proof mass acceleration noise, $\widetilde{x}_{\text{acc}}$, is correlated between different links that share the same proof mass. All other displacement noise contributions, combined in

$$\widetilde{x}_{\text{idp}} = \sqrt{\left(\langle\widetilde{x}_{\text{r/o}}^{\text{total}}\rangle_{\text{carrier}}\right)^2 + \left(\langle\widetilde{x}_{\text{r/o}}^{\text{total}}\rangle_{\text{sidebands}}\right)^2 + \left(\widetilde{x}_{\text{tml}}^{\text{total}}\right)^2 + \left(\widetilde{x}_{\text{ms}}^{\text{total}}\right)^2 + \left(\widetilde{x}_{\text{opn}}^{\text{total}}\right)^2}\,, \tag{54}$$

The amount of residual displacement noise due to laser frequency noise after TDI highly depends on **this** value hence we basically construct a virtual Michelson interferometer with an arm length difference equal to the ranging accuracy.

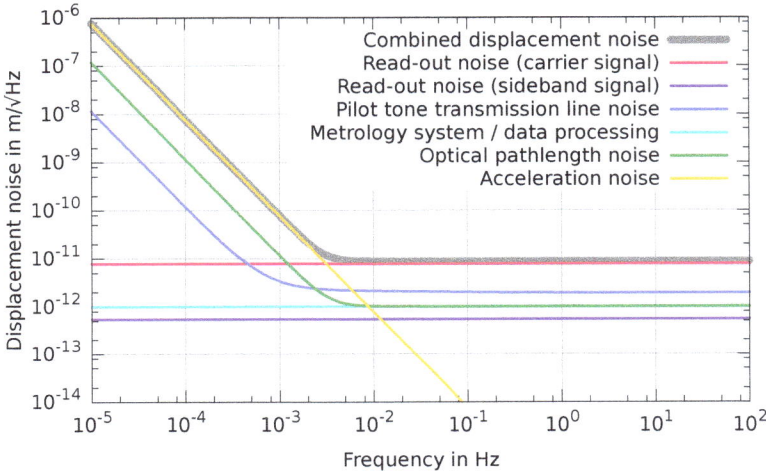

Figure 2.18: All effects that contribute to apparent displacement noise grouped into categories, and the resultant overall noise limit (combined displacement noise).

are independent between links. The total (single link) displacement noise which is used in all further evaluation of the observatory's sensitivity is given by

$$\widetilde{x}_{\text{total}} = \sqrt{(\widetilde{x}_{\text{acc}})^2 + (\widetilde{x}_{\text{idp}})^2}. \tag{55}$$

All parameters that were used to deduce the total displacement noise can be individually changed in the web application.

2.3 OBSERVATORY SENSITIVITY

There are a number of features of a gravitational wave observatory that are of astrophysical importance, one being the minimum characteristic gravitational strain amplitude that is detectable. In collaboration with Yan Wang [92] we will calculate the impact gravitational waves have on the observatory.

2.3.1 SINGLE LINK

To calculate the impact on one link when a gravitational wave passes though the observatory, we align the link with the unit vector e_x (in the direction of the x-axis) and consider a gravitational wave that propagates along a vector $k(\phi, \theta) = -(\cos\phi\cos\theta, \cos\phi\sin\theta, \sin\phi)$. The use of polar coordinates with latitude ϕ and longitude θ is illustrated in Figure 2.19. The oscillation of spacetime orthogonal to k happens along the orthogonal unit vectors $u(\phi, \theta)$ and $v(\phi, \theta)$ with x-axis components $u \cdot e_x = -\sin\phi\cos\theta$ and $v \cdot e_x = -\sin\theta$. Two influences have to be considered, both of them

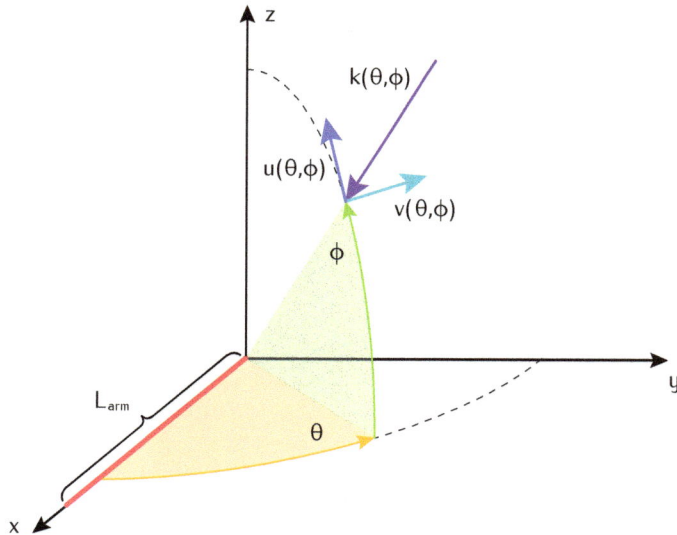

Figure 2.19: The response to gravitational waves of a single link (here: aligned with the x-axis) depends on the gravitational wave incident vector k with orthogonal components u and v. The actual oscillation is polarization dependent as indicated in Figure 1.8.

can reduce the impact of a gravitational wave of the link: the antenna pattern and the frequency response [93].

The antenna pattern $F(\theta, \phi)$ is a function of the sky position of the source (vector k) and combines the response for both polarization states. For a single link aligned with the x-axis it can be expressed by

$$F(\theta, \phi) = \frac{1}{2} \left[\underbrace{(u \cdot e_x)^2 - (v \cdot e_x)^2}_{+\text{ polarization}} + \underbrace{2(u \cdot e_x)(v \cdot e_x)}_{\times \text{ polarization}} \right] \tag{56}$$

$$= \frac{1}{2} \left(\sin^2 \phi \cos^2 \theta - \sin^2 \theta + 2 \sin \phi \cos \theta \sin \theta \right).$$

This function basically indicates which directions the gravitational wave observatory is sensitive to. The link will not be influenced by gravitational waves propagating along the x-axis at all. Independent of the polarization, the maximum impact can be observed for a +-polarized gravitational wave propagation orthogonal to the x-axis. A ×-polarized wave however does have no effect on the x-axis if propagating orthogonal to the x-axis. In general, laser interferometric gravitational wave observatories are sensitive to a very large fraction of the sky, hence they are usually referred to as omnidirectional detectors.

The frequency response $R(f, \theta, \phi)$ is a function of the gravitational wave frequency f, or—more accurately—the frequency of the influence of the gravitational wave propagating along vector k projected on the link vector x. It can be expressed by

$$R(f, \theta, \phi) = \frac{e^{2\pi i [1-kx] L_{\text{arm}}/\frac{c}{f}} - 1}{2\pi i [1-kx] L_{\text{arm}}/\frac{c}{f}} \times e^{-2\pi i k} \tag{57}$$

and depends on the actual arm length in relation to the wavelength of the gravitational wave $L_{\text{arm}}/\frac{c}{f}$ [92]. At low frequencies the frequency response is flat. For high frequencies, when the projected wavelength of the gravitational wave is similar to the arm length or smaller, the effect of the gravitational wave oscillation partially cancels out and the sensitivity is reduced.

Both influences combined give the total single link transfer function

$$T_{\text{link}}(f, \theta, \phi) = F(\theta, \phi) \times R(f, \theta, \phi) \tag{58}$$

and we can calculate its absolute average value over all sky positions ($\theta = 0 \ldots 2\pi$, $\phi = -\pi/2 \ldots \pi/2$)

$$T_{\text{link}}(f) = \sqrt{\left\langle |T(f, \theta, \phi)|^2 \right\rangle_{\text{sky}}}. \tag{59}$$

The effective strain sensitivity for a single link can now be formulated as the displacement noise over the single link transfer function

$$\sqrt{S_n(f)}_{\text{link}} = \frac{\widetilde{x}_{\text{total}}}{T_{\text{link}}(f) \times L_{\text{arm}}}. \tag{60}$$

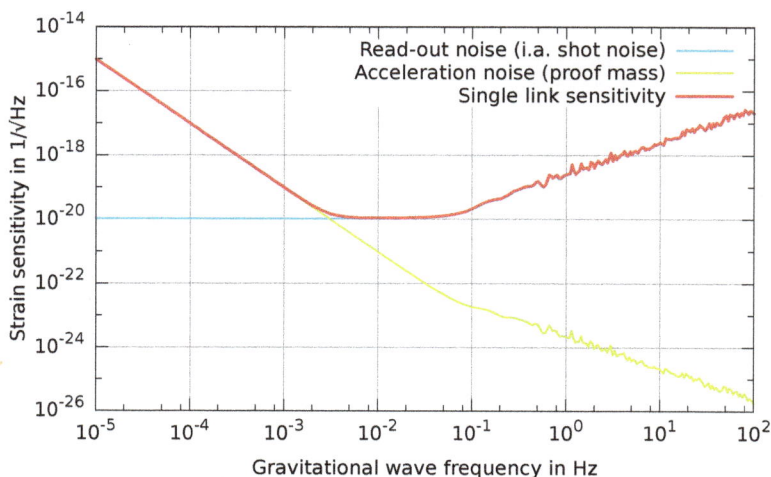

Figure 2.20: Single link strain sensitivity for the SAGA gravitational wave observatory compared to the individual noise contributions by carrier signal read-out noise and proof mass acceleration noise.

It is given in relative units (m/$\sqrt{\text{Hz}}$ per meter $= 1/\sqrt{\text{Hz}}$), thus the division by the arm length L_{arm}.

Figure 2.20 shows the effective single link strain sensitivity of the observatory specified above (red trace). Individual contributions by carrier signal read-out noise (blue) and proof mass acceleration noise (green) are shown. In a carefully designed observatory these two influences should limit the overall sensitivity.

The wiggles observable in the reduced sensitivity at high frequencies are the result of an attempt to reduce the response time of the web application— ideally below 400 ms, known as the Doherty threshold [94]—and the load on the web server performing the calculations. Thus we chose a time-saving averaging over only four values for $\theta = [0, 2, 4, 6]$ and four values for $\phi = [-1.41, 0.47, 0.47, 1.41]$. Yet this alone accounts for 16 different transfer functions with > 300 values each (50 values per frequency decade). For a perfect average over all sky positions the slope at high frequencies should become continuous.

2.3.2 FULL OBSERVATORY

The single link sensitivity is a good initial indicator of the observatory's performance. It can be used to compare different sets of parameters that share the same constellation to quickly identify limiting noise sources. This is the main purpose of the developed web application. In reality though, contributions like sideband signal read-out noise or pilot tone transmission chain noise have no effect when considering only one link. Instead, the sensitivity would be substantially reduced by frequency noise of the pre-stabilized lasers. Hence a single link cannot really be used to detect gravitational waves.

To calculate the actual sensitivity of the full observatory, we have to consider the combined responsivity of all links including their individual spatial orientation. This usually involves a full TDI simulation with realistic data streams precisely time-shifted (considering the ranging accuracy) to compensate for laser frequency noise. On top of that, all measurements have to by synchronized to a common reference (considering the pilot tone transmission fidelity) to remove clock noise from the individual measurements. This process is described in-depth by [88, 95, 96], but would require too many resources within the scope of the web application.

A good estimate of the full observatory sensitivity without excessive computational effort can be extrapolated from the single link sensitivity since in our case it already contains noise contributions due to limited ranging accuracy and pilot tone transmission fidelity. There are two effects: A) The combination of time-shifted signals results in an increased noise level: a thorough study of [96, 97] reveals that for a 60° virtual Michelson interferometer, TDI increases the proof mass acceleration noise at low frequencies by a factor of 4, while all other displacement noise contributions—which are uncorrelated between links—are increased by a factor of 2. B) The total number of virtual Michelson interferometers results in a general sensitivity improvement: a 3-arm triangular observatory can form three individual virtual Michelson interferometers, hence the overall sensitivity increases roughly by a factor of $\sqrt{3}$. Accordingly we can write the full observatory strain sensitivity approximately as

$$\sqrt{S_n\left(f\right)_{\mathrm{obs}}} \approx \frac{1}{\sqrt{3}} \times \frac{\sqrt{\left(4 \times \widetilde{x}_{\mathrm{acc}}\right)^2 + \left(2 \times \widetilde{x}_{\mathrm{idp}}\right)^2}}{T_{\mathrm{link}}\left(f\right) \times L_{\mathrm{arm}}} . \tag{61}$$

Figure 2.21 shows this full observatory sensitivity in red.

A two-arm observatory would be less sensitive by $\sqrt{3}$. Octahedral (12-arm) configurations use an enhanced post-processing technique called dis-

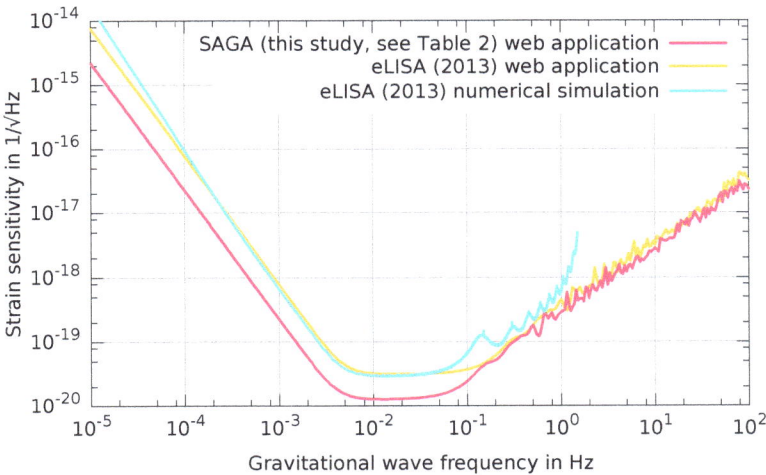

Figure 2.21: Approximate total strain sensitivity (all sky and polarization average) for the described eLISA-like observatory compared to a numerical TDI simulation for the eLISA (2013) gravitational wave observatory.

placement-noise free interferometry (DFI) [98] to suppress proof mass acceleration noise alongside any other spacecraft common mode displacement noise as well as laser frequency noise. For these cases the web application uses the approximations described in [62].

For comparison, a numerical TDI simulation that was done for the eLISA (2013) gravitational wave observatory mission concept is shown in blue. This study was part of 'The Gravitational Universe' White Paper [51]. eLISA (2013) used slightly different parameters, namely only 4 links, smaller arm length, telescope diameter and heterodyne frequency, and higher laser power. A list of important parameters that differ from the ones of SAGA can be found in Table 3.

Table 3: Important parameters that differ from Table 2 (SAGA) to correspond to the parameter set used for the eLISA (2013) mission study. As a result, eLISA (2013) will experience a lower carrier read-out noise but still suffer from a reduced sensitivity due to the shorter arms.

Parameter		Value
Number of links	$N_{\text{links}} =$	4
Average arm length	$L_{\text{arm}} =$	1 000 000 km
Heterodyne frequency (max.)	$f_{\text{het}} =$	12 MHz
Optical power (to telescope)	$P_{\text{tel}} =$	2 W
Telescope diameter	$d_{\text{tel}} =$	20 cm

The result from the web application for this new parameter set with the full observatory strain sensitivity approximated by Equation 61 is shown in orange. Although for this approximated sensitivity the wiggles at high frequencies are again due to a sloppy averaging. Similar wiggles in the sensitivity deduced by the numerical simulation are a real consequence of the TDI algorithms. This shows the limitations of our approximation. Nevertheless it is sufficient for the purpose of parameter optimization and is a very close match to the real sensitivity. Thus we can use it to investigate the astrophysical relevance of the observatory.

2.3.3 DETECTION LIMIT

The scientific value of an observatory is related to the number and type of sources it can detect. In Figure 2.22 we use all parameters from Table 2 to plot the observatory's detection limit

$$ h_{\text{c}}\left(f\right) = \sqrt{f} \times \sqrt{S_n\left(f\right)_{\text{obs}}} \tag{62} $$

where the signal-to-noise ratio equals 1 [51, p. 14]. We can compare this to the characteristic gravitational wave strain amplitudes (given in m/m) for selected gravitational wave sources as described in Section 1.2.1. For quasi-monochromatic sources the accumulated signal after one year of observation time is given. Amplitudes of all other broadband sources are plotted as is, although their actual SNR can be higher due to matched filtering techniques during data analysis.

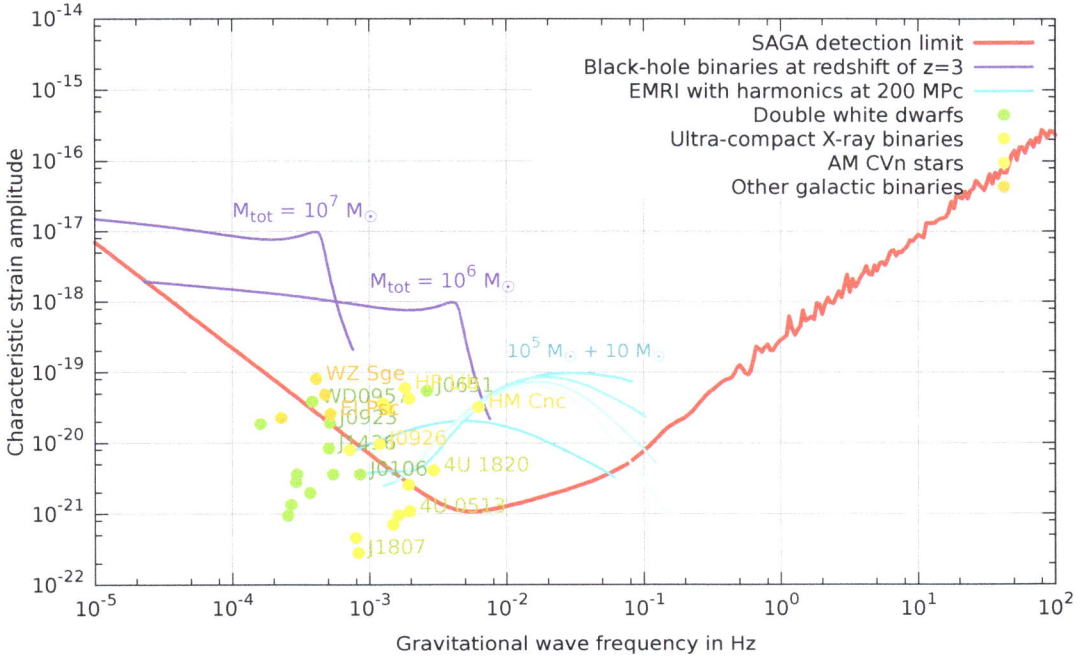

The exact same quantities were used for Figure 1.11. Plots generated by the web application can be used to compare the detection limit between a variety of mission concepts and bring them face to face with ground based detector sensitivities.

2.4 WEB APPLICATION

All of the calculations above can be performed and documented for any specific set of mission parameters by the "Gravitational Wave Observatory Designer". This web application—which is publicly available on the Internet—was developed in the context of this thesis. It provides an HTML5 based graphical user interface (GUI) designed with jQuery, a cross-platform JavaScript library, and Elements from Polymer, an open-source Web Components-based library made available by *Google Inc.* Although only Chrome (and other Blink-based browsers like Opera) ship with native platform support for Web Components, a JavaScript foundation layer provides compatibility for the latest version of all 'evergreen' (self updating) web browsers. That currently includes Chrome (also Android and Canary versions), Firefox, Internet Explorer (version 10 and up), and Safari (version 6 and up, also mobile versions).

The compliance with *Google Inc.*'s 'Material Design' guidelines allows for a unified user experience across a wide range of devices, screen sizes, and formats. Examples are shown in Figure 2.23.

Figure 2.22: Observatory detection limit (for $SNR = 1$) and dimensionless characteristic strain amplitudes for different gravitational wave sources. Two traces for systems of binary black holes at redshift of $z = 3$ (total mass $M_{tot} = 10^7 M_\odot$ and $= 10^6 M_\odot$), where the former trace starts at low frequencies ≈ 1 month, the latter ≈ 1 year before the plunge (spike in the trace). First 5 harmonics of one eccentric Extreme Mass Ratio Inspiral (EMRI) for an object with mass $m = 10 M_\odot$ captured by a massive black hole of mass $M = 10^5 M_\odot$ at 200 Mpc distance. The EMRI trace starts at low frequencies many years before the merger. A selection of known ultra-compact binary stars (dots) for 1 year of observation time is also shown.

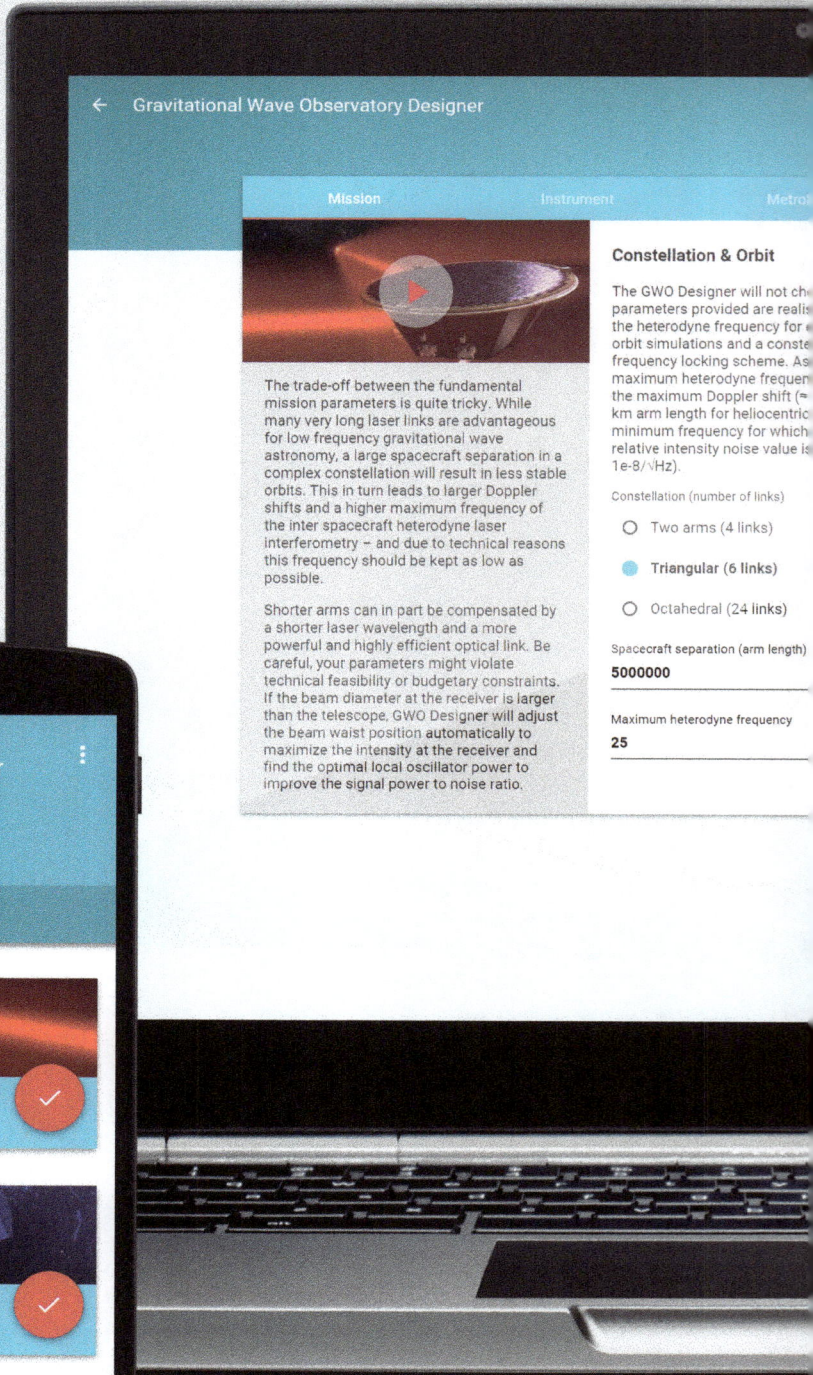

Gravitational Wave Observatory Designer

Mission   Instrument   Metro...

Constellation & Orbit

The GWO Designer will not ch...
parameters provided are realis...
the heterodyne frequency for ...
orbit simulations and a conste...
frequency locking scheme. As ...
maximum heterodyne frequenc...
the maximum Doppler shift (≈ ...
km arm length for heliocentric ...
minimum frequency for which ...
relative intensity noise value is...
1e-8/√Hz).

Constellation (number of links)

○ Two arms (4 links)

● Triangular (6 links)

○ Octahedral (24 links)

Spacecraft separation (arm length)
5000000

Maximum heterodyne frequency
25

The trade-off between the fundamental
mission parameters is quite tricky. While
many very long laser links are advantageous
for low frequency gravitational wave
astronomy, a large spacecraft separation in a
complex constellation will result in less stable
orbits. This in turn leads to larger Doppler
shifts and a higher maximum frequency of
the inter spacecraft heterodyne laser
interferometry – and due to technical reasons
this frequency should be kept as low as
possible.

Shorter arms can in part be compensated by
a shorter laser wavelength and a more
powerful and highly efficient optical link. Be
careful, your parameters might violate
technical feasibility or budgetary constraints.
If the beam diameter at the receiver is larger
than the telescope, GWO Designer will adjust
the beam waist position automatically to
maximize the intensity at the receiver and
find the optimal local oscillator power to
improve the signal power to noise ratio.

**Gravitational Wave
Observatory Designer**

Choose a Preset

Classic LISA

eLISA (2013)

Figure 2.23: Graphical user interface of the "Gravitational Wave Observatory Designer": The compliance with Google Inc.'s 'Material Design' guidelines allows for a uni ed user experience across a wide range of devices, screen sizes, and formats.

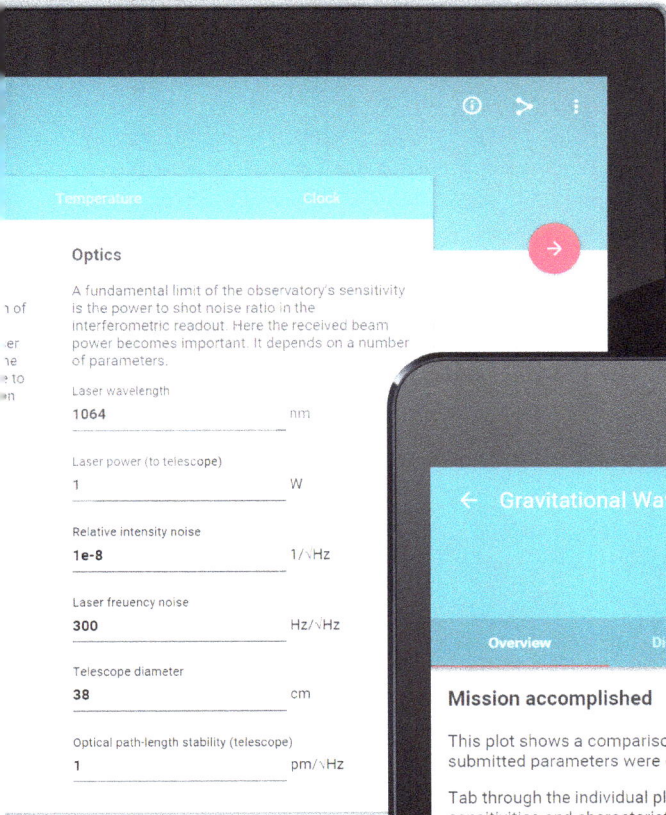

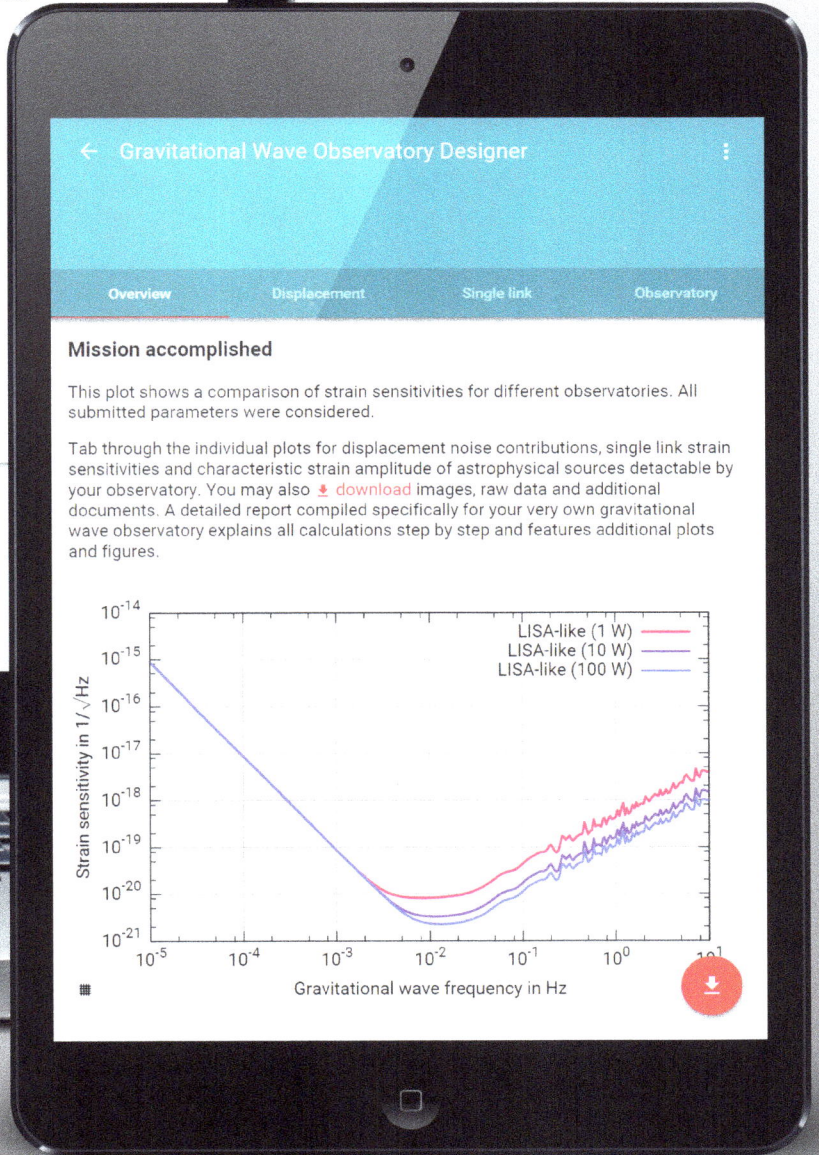

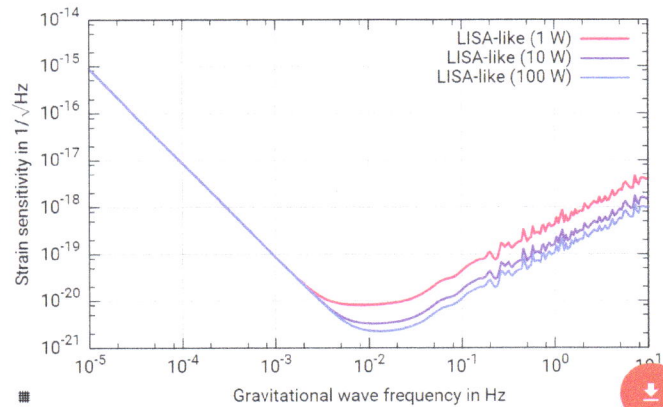

Temperature   Clock

Optics

A fundamental limit of the observatory's sensitivity is the power to shot noise ratio in the interferometric readout. Here the received beam power becomes important. It depends on a number of parameters.

Laser wavelength
1064   nm

Laser power (to telescope)
1   W

Relative intensity noise
1e-8 $1/\sqrt{Hz}$

Laser freuency noise
300                                 $Hz/\sqrt{Hz}$

Telescope diameter
38   cm

Optical path-length stability (telescope)
1                                   $pm/\sqrt{Hz}$

← Gravitational Wave Observatory Designer

Overview   Displacement   Single link   Observatory

Mission accomplished

This plot shows a comparison of strain sensitivities for different observatories. All submitted parameters were considered.

Tab through the individual plots for displacement noise contributions, single link strain sensitivities and characteristic strain amplitude of astrophysical sources detactable by your observatory. You may also ⬇ download images, raw data and additional documents. A detailed report compiled specifically for your very own gravitational wave observatory explains all calculations step by step and features additional plots and figures.

2.4.1 BACK END

All calculations are done by a Perl CGI back end that is connected to the GUI via Ajax, a technique for asynchronous client-side JavaScript and XML. It utilizes Perl modules such as Math::Cephes, PDL, and Math::Complex, and interfaces with gnuplot, an open source command-line program to generate graphics in various formats including interactive SVG plots. PDF documents are created by LaTeX, a document preparation system and markup language, and the raw data is also available for download in ZIP archive file format. Results for different designs can be compared easily as parameters can be given as arrays. We also provide default parameter sets for some known design studies. Furthermore, parameter sets previously processed can be restored by a recovery mechanism.

2.4.2 FRONT END

An instructional video for the **web application** is available that will guide you through the most important features.

simonbarke.com/phd/god

The **web application** is publicly available at http://spacegravity.org/designer. Let me quickly guide you through the features of the Gravitational Wave Observatory Designer and show you some nifty tricks.

"When you open the web application, you are presented with three options. You can start from scratch with completely blank parameter specifications, or provide a recovery code to modify an earlier design. New users should choose the 'Load mission preset' option. Here you can choose between 'OGO', an experimental yet quite interesting design study [62], 'eLISA' [99], and 'Classic LISA'[100]. Your choice will prepopulate all parameter specifications. Maybe most important are the constellation and orbit specifications. The web application is capable of calculating the sensitivity for constellations with 2, 3, and 12 arms.

All parameters can be given as integer, decimal value, or in scientific notation. You can even enter an array of numbers, separated by commas. When you do that for multiple parameters, each possible combination will be calculated separately. You can provide details for the laser systems and optical telescopes, for the optical bench and the photo receiver electronics, the phase measurement system and the gravitational reference sensors, enter values for the electrical and optical transmission lines, provide properties of the signal modulation, and specify detailed temperature noise levels. The web application will optimize parameters like the waist radius of the laser beams, calculate the magnitude of many different noise sources, and determine the best possible sensitivity for your very own gravitational wave observatory.

You can download the final observatory strain sensitivity compared to other mission studies, get a break down of all the different noise contributions to easily identify the influences limiting your sensitivity, and compare the observatory's strain sensitivity with the usually dominating influences of read-out noise and acceleration noise. The characteristic strain amplitude of your observatory is given in relation to typical astrophysical sources of gravitational waves. In each plot you can select items in the key to toggle the visibility of the related traces. So if a plot becomes too crowded, just hide some of the items. You can also click in the plot to display X and Y coordinates for any point. Of course you may use all plots and data under Creative Commons Attribution license.

If you need help with a certain parameter, you can always switch on the tool tips in the action bar. Pictures, explanations, or standard values will then be shown when you hover over a parameter. You may save your work at any point. A recovery code will be displayed that you can use to continue with your design."

You can directly work with the parameters of the SAGA concept used in this chapter. Simply visit http://spacegravity.org/designer and enter recovery code 'bd16-cce5-5a7d'.

Gravitational Wave Observatory Designer SAGA preset.

spacegravity.org/designer/
#rc=bd16-cce5-5a7d

2.4.3 LIMITATIONS

The present web application was developed to quickly identify limiting noise sources common to all laser interferometric gravitational wave observatories. Noise contributions addressed in this chapter are not intended to be exhaustive. Additional systems specific to the detailed observatory design might add a significant amount of excess noise. Also for most contributions white noise was assumed, however, in reality the noise shapes will be more complex. Future updates may include additional noise contributions and individual noise shapes.

Nevertheless, the "Gravitational Wave Observatory Designer" is the most comprehensive sensitivity curve generator for a wide range of spaceborne gravitational wave detectors to my knowledge. It will educate and inspire on the subject of interferometric gravitational wave observatories, quickly show the potential and limitations of new ideas and concepts, and help to explore the parameter space in preparation for the planned call for mission concepts for ESA's L3 mission opportunity, expected in 2016 [101].

Many parameter combinations require that their feasibility is assessed in a separate detailed study. The maximum heterodyne frequency for example depends on many different factors, some of technical nature, some driven by gravitational disturbances within our solar system. Its exact value has to be carefully evaluated for each individual case. ∎

3

For the Classic LISA mission, the expected relative velocity along the line-of-sight between any two spacecraft of up to $\Delta v = 20\,\mathrm{m\,s^{-1}}$ leads to a maximum Doppler shift of roughly 19 MHz according to Equation 8. Thus for a long time it was assumed that the heterodyne frequency range of Classic LISA could be easily kept between 2. . . 20 MHz [102]. In reality, the situation is much more complicated, although this fact has never been conclusively addressed as of this writing.

To avoid any zero crossings and allow for stable read-out electronics a 0. . . 2 MHz gap was chosen.

3.1 SHOT NOISE LIMITS OF DIFFERENT CONCEPTS

In this chapter, I determine the heterodyne frequency range for all three mission concepts, Classic LISA, eLISA (2013), and SAGA. Operating the observatory at the minimum or maximum of this range can potentially deteriorate the observatory's sensitivity. I want to make sure that this is not the case. Since all mission concepts were designed to be limited by shot noise in the carrier signal read-out, we need to keep all influences that depend on the actual read-out frequency well below this level. The influence of shot noise is different though for the particular concepts. Thus I want to introduce all three mission concepts in a little more detail.

Gravitational Wave Observatory Designer SAGA preset with correct heterodyne and modulation frequencies.

spacegravity.org/designer/ #rc=f5d1–caa1–5ace

3.1.1 SAGA

The SAGA concept maybe comes closest to how the first LISA-like mission will look like. It was dealt with in depth in Chapter 2. The important parameters for now are the arm length ($L_{arm} = 2\,000\,000$ km), the laser power

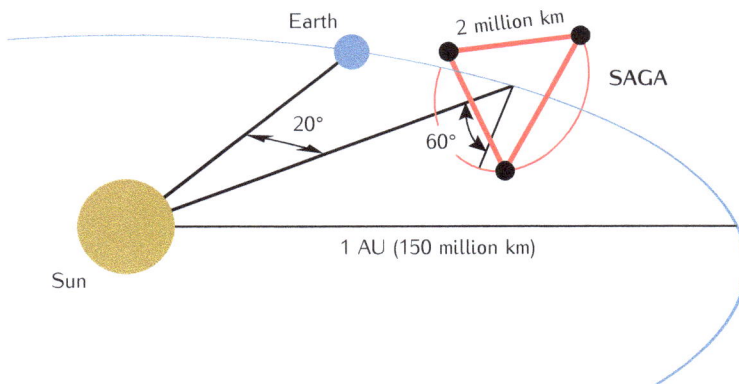

Figure 3.1: Each SAGA spacecraft follows an Earth-trailing heliocentric orbit with an inclination of about 1° with respect to the ecliptic. This results in a stable triangular formation inclined by 60° with respect to the ecliptic which is kept at a constant 20° behind Earth.

delivered to the telescope ($P_{\text{tel}} = 1.65\,\text{W}$) and the diameter of the optical telescopes ($d_{\text{tel}} = 26\,\text{cm}$). These parameters are in line with recent studies e.g. by *Airbus Defence and Space* (formerly *EADS Astrium*). Allowing for 7.5% of the carrier power in each auxiliary clock sideband, Equation 32 shows that the equivalent displacement noise due to shot noise in the carrier signal is $\left\langle \tilde{x}^{\text{sn}}_{\text{r/o}} \right\rangle_{\text{carrier}} = 6.58 \times 10^{-12}\,\frac{\text{m}}{\sqrt{\text{Hz}}}$. Furthermore, this concept uses a total of $N_{\text{links}} = 6$ individual laser links for improved sensitivity, instantaneous polarization discrimination, better spatial resolution, and higher redundancy. Each spacecraft follows an Earth-trailing heliocentric orbit with an inclination of about $1°$ with respect to the ecliptic as illustrated in Figure 3.1. This results in a stable triangular formation, inclined by $60°$ with respect to the ecliptic, which is kept at a constant $20°$ behind Earth.

The SAGA concept sits somewhere in the middle between eLISA and Classic LISA. While eLISA provides a lower shot noise, Classic LISA yields a better detection limit due to its longer arms.

3.1.2 ELISA (2013)

The main differences between SAGA and the 2013 incarnation of eLISA [99] were already summarized briefly in Table 3 on page 58. The latter features an arm length of $L_{\text{arm}} = 1\,000\,000\,\text{km}$ and a laser power of $P_{\text{tel}} = 2.00\,\text{W}$. The shorter arm length allows for smaller optical telescopes of $d_{\text{tel}} = 20\,\text{cm}$ diameter and an overall more compact spacecraft design. This leads to a carrier signal shot noise of $\left\langle \tilde{x}^{\text{sn}}_{\text{r/o}} \right\rangle_{\text{carrier}} = 5.05 \times 10^{-12}\,\frac{\text{m}}{\sqrt{\text{Hz}}}$. The concept includes only $N_{\text{links}} = 4$ laser links to reduce the mission cost. As indicated in Figure 3.2, a fuel saving orbit transfer strategy was chosen that results in an Earth-trailing heliocentric runaway orbit that starts at $10°$ behind Earth and slowly drifts away to $30°$ within the 2 years of nominal mission lifetime. Such orbits would limit the maximum mission lifetime as the observatory slowly moves out of communication range with Earth within a timespan that may be shorter than the actual lifetime of the spacecraft.

Figure 3.2: Each eLISA spacecraft follows an Earth-trailing heliocentric orbit with an inclination of about $1°$ with respect to the ecliptic. This results in a stable triangular formation inclined by $60°$ with respect to the ecliptic. The original concept planned for a runaway orbit that slowly drifts away, starting at $10°$ behind Earth.

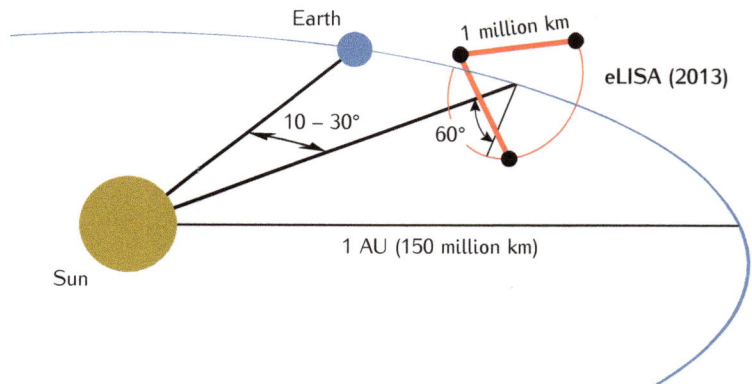

3.1.3 CLASSIC LISA

The Classic LISA mission [100] is the by far largest viable gravitational wave observatory concept to date. With $L_{arm} = 5\,000\,000$ km long arms and a laser power of $P_{tel} = 1.00$ W it houses telescopes $d_{tel} = 38$ cm in diameter. The carrier signal shot noise of $\langle \tilde{x}^{sn}_{r/o} \rangle_{carrier} = 9.90 \times 10^{-12} \frac{m}{\sqrt{Hz}}$ is significantly higher compared to other mission concepts, a fact that is compensated by the long arms and the $N_{links} = 6$ laser links which, combined, result in a best-in-class sensitivity. It is maybe the most sophisticated mission concept, planned to be located $20°$ behind Earth in a stable heliocentric orbit as shown in Figure 3.3. This location offers a good trade-off between low gravitational disturbances and good communication range. This configuration had been the baseline for more than a decade and has been thoroughly investigated by, e.g., [75] and industrial studies [103].

Gravitational Wave Observatory Designer Classic LISA preset with correct heterodyne and modulation frequencies.

spacegravity.org/designer/ #rc=009c-f78b-5fe4

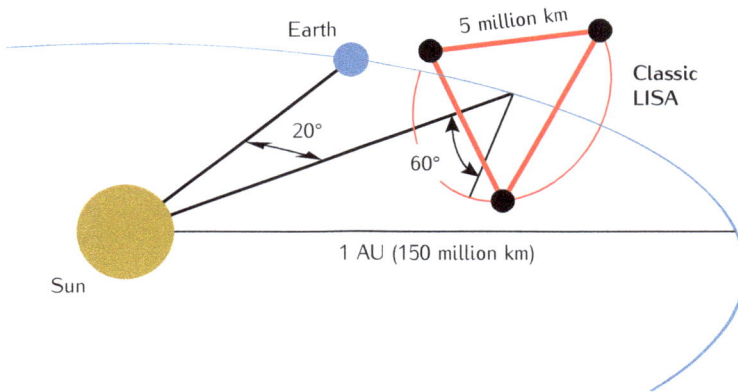

Figure 3.3: Each LISA spacecraft follows an Earth-trailing heliocentric orbit with an inclination of about 1° with respect to the ecliptic. This results in a stable triangular formation inclined by 60° with respect to the ecliptic which is kept at a constant 20° behind Earth.

Table 4 compares mission parameters relevant for the different shot noise levels for the three concepts. To get a better understanding of the actual dimensions, Figure 3.4 impressively shows how huge and how far away gravitational waves observatories are in a to-scale illustration of the inner solar system. In Figure 3.5 the Sun and inner planet diameters as well as the lunar orbit to scale next to eLISA (2013) are drawn. In a way, gravitational wave observatories will be the largest instruments ever constructed by humankind.

Parameter		eLISA (2013)	SAGA	Classic LISA	
Number of links	$N_{links} =$	4	6	6	
Average arm length	$L_{arm} =$	1 000 000	2 000 000	5 000 000	km
Optical power	$P_{tel} =$	2.00	1.65	1.00	W
Telescope diameter	$d_{tel} =$	20	26	38	cm
Received laser power	$P_{rec} =$	994	586	259	pW
Carrier shot noise	$\langle \tilde{x}^{sn}_{r/o} \rangle_{carrier} =$	5.05×10^{-12}	6.58×10^{-12}	9.90×10^{-12}	$\frac{m}{\sqrt{Hz}}$

Table 4: Main differences in mission parameters between eLISA (2013), SAGA and Classic LISA that are relevant for the level of shot noise in the signal read-out.

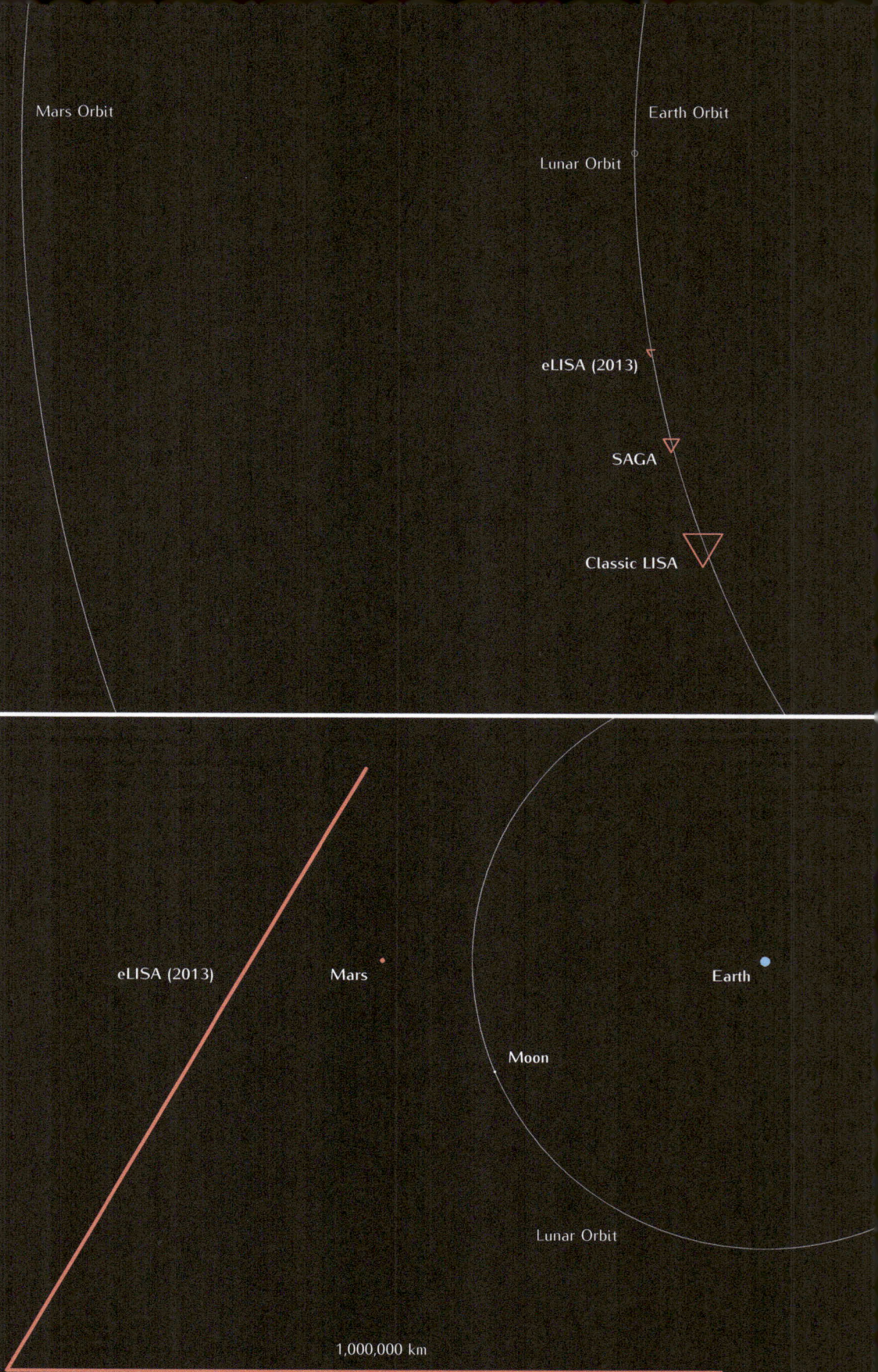

Mars Orbit

Earth Orbit

Lunar Orbit

eLISA (2013)

SAGA

Classic LISA

eLISA (2013)

Mars

Earth

Moon

Lunar Orbit

1,000,000 km

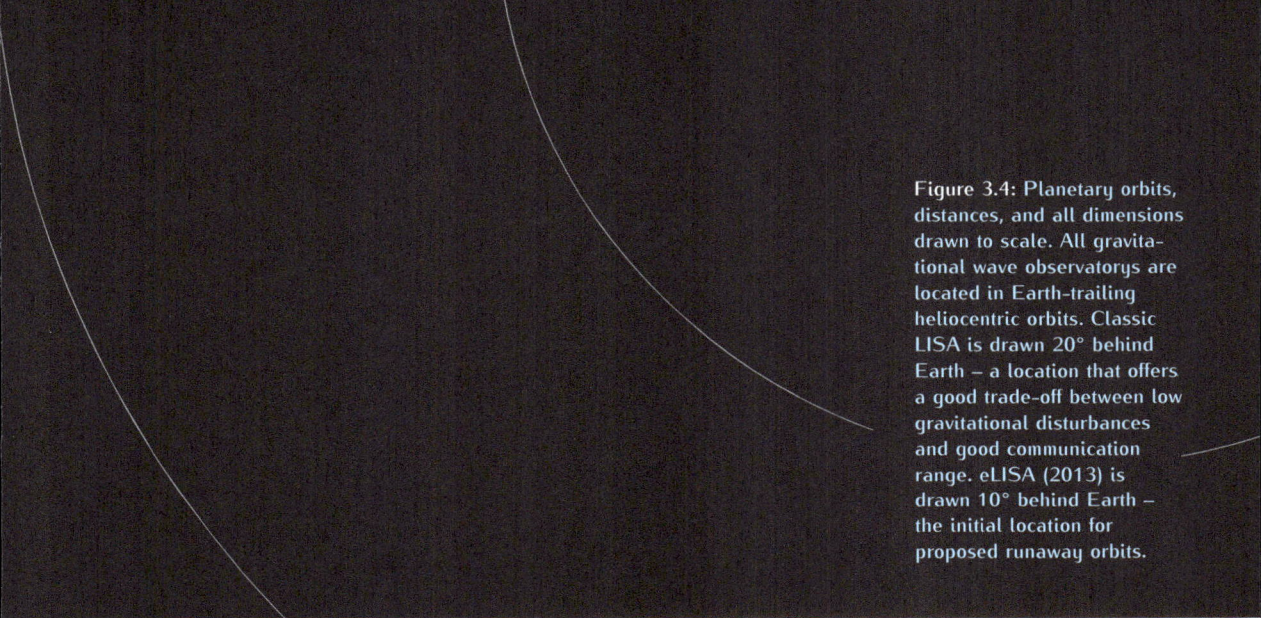

Figure 3.4: Planetary orbits, distances, and all dimensions drawn to scale. All gravitational wave observatorys are located in Earth-trailing heliocentric orbits. Classic LISA is drawn 20° behind Earth – a location that offers a good trade-off between low gravitational disturbances and good communication range. eLISA (2013) is drawn 10° behind Earth – the initial location for proposed runaway orbits.

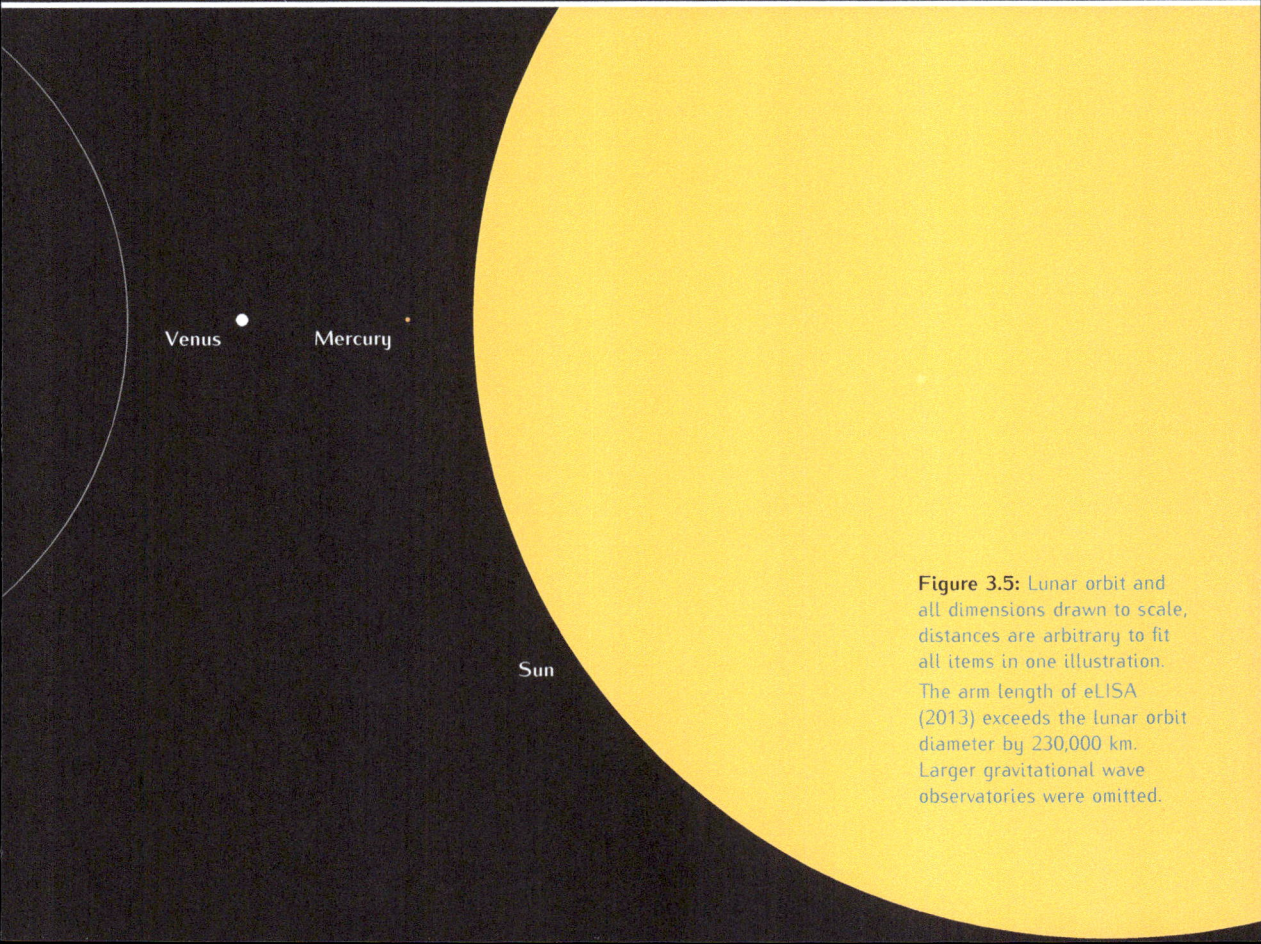

Venus

Mercury

Sun

Figure 3.5: Lunar orbit and all dimensions drawn to scale, distances are arbitrary to fit all items in one illustration.

The arm length of eLISA (2013) exceeds the lunar orbit diameter by 230,000 km. Larger gravitational wave observatories were omitted.

3.2 FORBIDDEN FREQUENCY DOMAINS

There are a number of frequencies and frequency ranges that have to be avoided. Harmonics of the pilot tone added to the heterodyne signal read-out for example may interfere with the carrier and sideband beat notes described in Section 2.2.2. This naturally depends on the specific pilot tone frequency and the purity of the signal. A more general issue is the relative intensity noise (RIN) of lasers at low Fourier frequencies. Power fluctuations at a certain frequency cannot be distinguished from the heterodyne signal at the same frequency. Hence RIN deteriorates the signal-to-noise ratio of interferometric length measurements at a given heterodyne frequency.

3.2.1 LASER RELATIVE INTENSITY NOISE (RIN)

Using Equation 25 the frequency-dependent influence of laser relative intensity noise in the signal read-out given as equivalent displacement noise can be written as

$$\left\langle \tilde{x}_{\text{r/o}}^{\text{rin}} \right\rangle (f) = \frac{\lambda_{\text{laser}}}{2\pi} \frac{1}{J_0(m)^2} \times RIN(f) \sqrt{\frac{P_{\text{local}}^2 + P_{\text{rec}}^2}{2\eta_{\text{het}} P_{\text{local}} P_{\text{rec}}}} \tag{63}$$

where f denotes the heterodyne frequency, or, more precisely, the frequency at which the actual read-out is performed. Usually a higher heterodyne frequency increases the influence of many related noise sources. In this case, however, measurements reveal that the relative intensity noise increases dramatically at low frequencies.

Figure 3.6 shows as an example the RIN of a 1064 nm nonplanar ring oscillator (NPRO) laser (model "Mephisto 500 NE spezial" by *InnoLight GmbH*, now *Coherent Inc.*, serial number: 1915A, laser diode current = 0.8 A, temperature = 29 °C) at 500 mW of optical output power. Similar laser systems are strong candidates to be used as master laser on-board a future gravitational wave observatory.

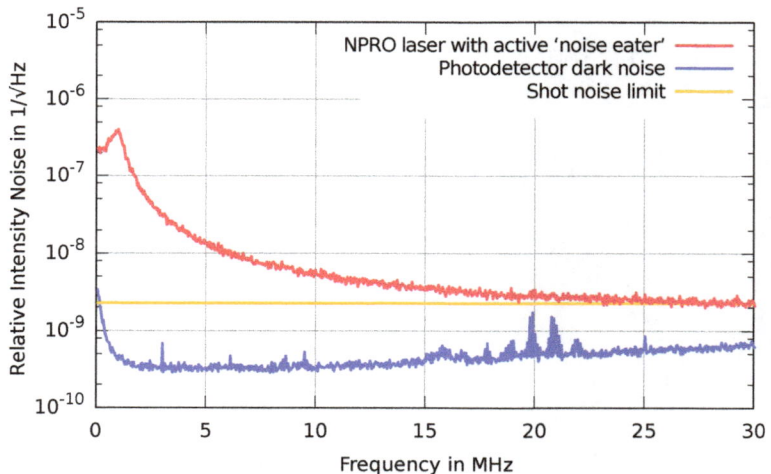

Figure 3.6: RIN of a 1064 nm nonplanar ring oscillator (NPRO) laser: NPRO with active 'noise eater' (red), dark noise (blue), shot noise limit (yellow). The shot noise limits the RIN for frequencies aove 20 MHz.

The laser system was operated with active 'noise eater', a build-in noise suppression technique designed to suppress a usually sharp power modulation peak around 1 MHz. This peak is still visible in the measurement (red trace), though substantially attenuated. Measurements were performed using a photo detector build for the 'LIGO Laser Diagnostic Breadboard' [104] following a procedure described in [105]. Dark noise of the photo detector (blue) was found to be significantly below the relative intensity noise level of the laser. For an incident light power of 175 mW an average voltage level of $U_{avg} = 12.9\,V$ was measured at the photo detector. The relative shot noise floor then calculates as

$$N_{sn} = \frac{\sqrt{2\,q_e\,I_{dc}}}{I_{dc}} = \sqrt{2\,q_e\,200\,\Omega/U_{avg}} = 2.23 \times 10^{-9}\,/\sqrt{Hz} \quad (64)$$

The correlation between incident light power and average voltage level at the LF output of the photo detector was measured to be 13.5 mW per volt.

(shown in yellow) and turns out to be the limiting factor of the relative intensity noise for frequencies above 20 MHz.

These measurements are not directly representative for the expected RIN in the actual space mission. Here, a laser amplifier stage is required to boost the power to the required 1 . . . 2 watts range. This stage might add additional relative intensity noise or may introduce new limitations. Representatively I measured the excess relative intensity noise of a commercial polarization maintaining fiber amplifier (model "PSFA-1064-01-10W-2-3" by *Nufern*). This particular system provided a power output range of 1 . . . 10 W for a variable seed power of 1 . . . 15 mW.

Many core- and cladding-pumped Ytterbium doped fiber amplifiers were tested at breadboard level to assess the technology readiness [66, 106, 107]. There is no fully compatible of-the-shelf laser system available to date.

The measurement setup is illustrated in Figure 3.7. The same NPRO laser was used to seed the fiber amplifier after attenuating it by a $\lambda/2$ wave plate and a polarizing beam splitter (PBS) to the required power levels of 2 . . . 15 mW. The output power of the amplifier stage was internally measured by a power monitor built into the control unit of the amplifier ('Nufern Control') and verified with thermal head powermeter (model 'ITDH-100P' by *Lasermet*). The results of the individual measurements differed by up to 17% as can be seen from the red and green bars in Figure 3.8. The photo detector of the LIGO Laser Diagnostic Breadboard mentioned above was used to measure the fre-

Figure 3.7: Measurement setup for RIN measurements of NRPO laser and fiber amplifier. A powermeter behind a flipping mirror was installed to verify the fiber amplifier output power. Only a fraction of the light was used for the actual RIN measurement.

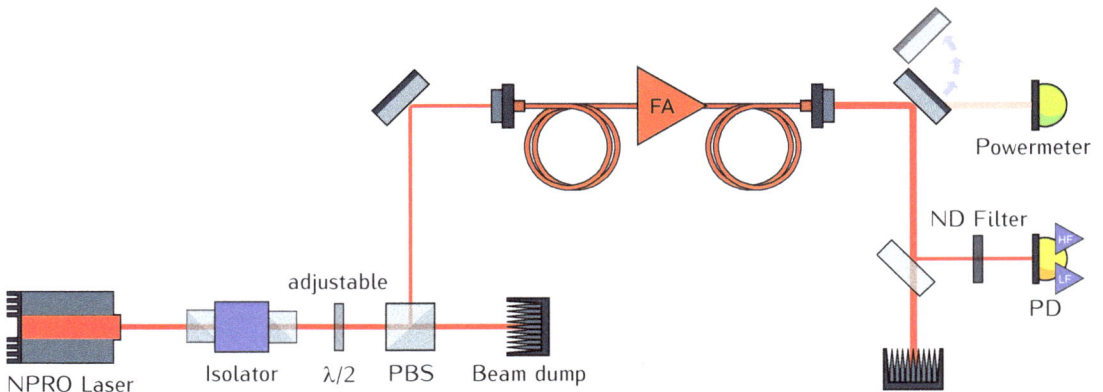

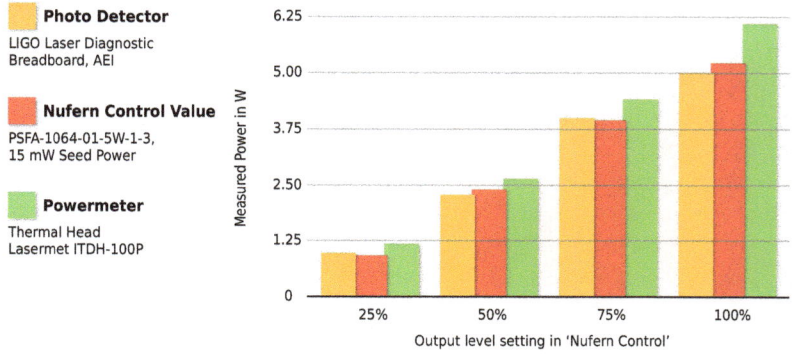

The power values of the photo detector and the internal power monitor suggest that the actual power might be significantly higher. The used **calibration value** thus is very conservative and the real RIN could be lower by up to 17%.

quency resolved relative intensity noise. Due to the nonlinearities and generally unreliable values of both power measuring tools, we assumed a real output power of 5 W as stated in the data sheet of the amplifier for a 100% output level setting in the Nufern control unit. At this setting, the DC output of the photo detector returned a voltage of 13.36 V. I used a calibration value of 2.7 W/V accordingly. Only a fraction of the light intensity split off by an uncoated glass plate and further attenuated by a subsequent neutral density filter was used for this measurement. The attenuation stage remained unaltered throughout all RIN measurements.

In Figure 3.9 the relative intensity noise is plotted over Fourier frequency for four different fiber amplifier output power settings (25%, 50%, 75%, 100%) with corresponding internal power monitor values between 0.91 and 5.31 W. Two different seed powers were used (2 mW, yellow and green traces, and 14.5 mW, cyan and blue traces). It can readily be seen that there is no correlation between the RIN and output power, and the measurements were not limited by the actual relative shot noise level. The strong correlation between RIN noise floor and seed power hints at a limitation by amplified seed laser shot noise.

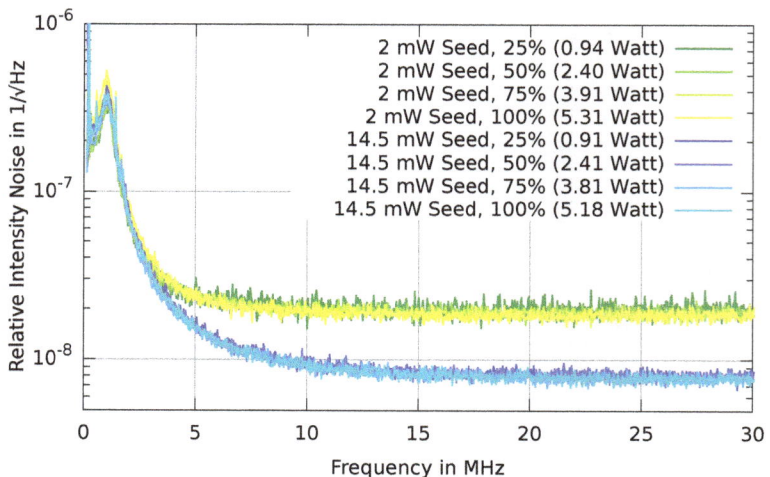

Figure 3.9: RIN plotted over Fourier frequency for four different fiber amplifier output power settings and two different seed powers. There is no correlation between the RIN and output power, but all the more between RIN and seed power.

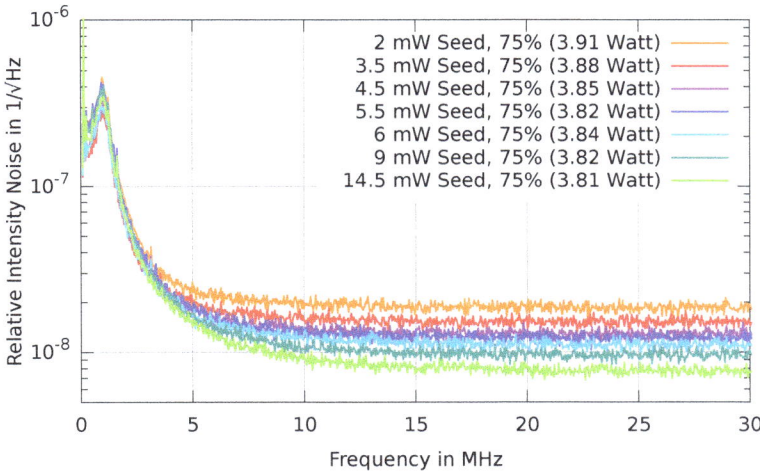

Figure 3.10: RIN plotted over Fourier frequency for seven different seed powers at a constant fiber amplifier output power. A correlation between RIN noise floor and seed power is clearly visible.

To have a better understanding of the correlation, the same measurement was performed at a constant fiber amplifier output power of $\approx 3.8\,\mathrm{W}$ (75%) for seven different seed powers (2.0, 3.5, 4.5, 5.5, 6.0, 9.0, 14.5 mW). Results are shown in Figure 3.10. A detailed analysis shows a non-linear relationship between seed power and white power noise floor level RIN_{floor} that scales with the inverse square-root of the seed power P_{seed} (in mW) as $RIN_{\mathrm{floor}} = 1.4 \times 10^{-9} + 2.5 \times 10^{-8}/\sqrt{P_{\mathrm{seed}}}$. An actual fit to the measured data is presented in Figure 3.11 (blue trace). It matches the corresponding relative shot noise levels for the low power seed laser (yellow) quite well.

We find, that the RIN of current fiber amplifiers is fundamentally limited by the shot noise in the seed laser. To keep the relative intensity noise at a lower level, a high seed power is desirable. I will use the lowest trace from Figure 3.10 in all further considerations. This is a quite conservative assumption considering that in the final mission, seed powers > 20 mW might easily be possible.

For the used seed powers **shot noise levels** are 2.08×10^{-8}, 1.57×10^{-8}, 1.39×10^{-8}, 1.25×10^{-8}, 1.20×10^{-8}, 9.80×10^{-9}, and $7.72 \times 10^{-9}/\sqrt{\mathrm{Hz}}$.

The **seed power** is limited by the damage threshold of the electro-optic modulator (see Section 5.1.2) and its attenuation since this component will be placed directly in front of the fiber amplifier.

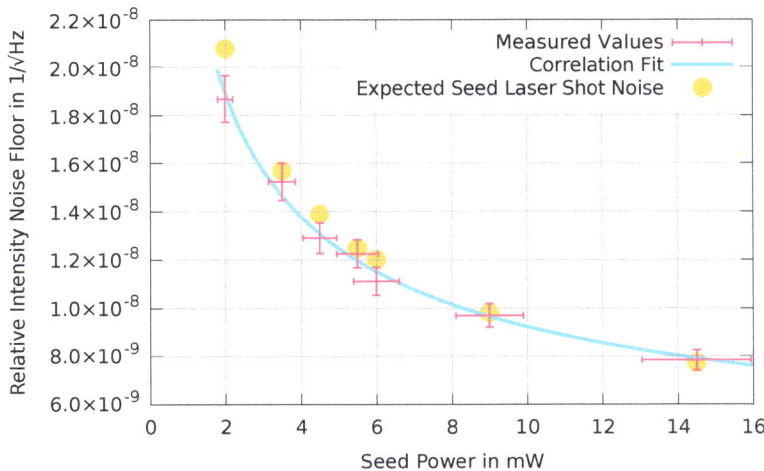

Figure 3.11: It could be found that the white power noise floor level RIN_{floor} scales with the inverse square-root of the seed power P_{seed} (in mW) as $RIN_{\mathrm{floor}} = 1.4 \times 10^{-9} + 2.5 \times 10^{-8}/\sqrt{P_{\mathrm{seed}}}$.

3.2.2 INFLUENCE OF RIN FOR DIFFERENT MISSION CONCEPTS

There is a complex correlation between the properties of the photoreceiver electronics (photodiode and transimpedance amplifier, see Section 2.1.3), the actual read-out scheme, the level of relative intensity noise, the received laser power and heterodyne efficiency, and the optimal local laser power used to create the carrier beat note. In the following I will use all parameters as stated in Table 2 on page page 29 with individual adaptations from Table 4 on page page 67 according to the specific mission concept.

We now have to find the level of relative intensity noise with a corresponding optimal local laser power (see Section 2.2.1.4) which results in a RIN-induced read-out noise that is below the carrier shot noise by a certain factor. Building up on the work already done in Section 2.2.1 this is easy now. We can use Equation 28 for the electronic noise—corrected by the carrier read-out factor $1/J_0(m)^2$—to find the local oscillator power which results in a certain carrier read-out noise level. Ideally, this level is an order of magnitude below the corresponding carrier shot noise stated in Table 4. As obvious from Figure 2.10 on page page 41, local oscillator power is optimal if the RIN induced read-out noise (calculated by Equation 63) for the same power yields the exact same noise level.

Table 5 shows a summary of optimal local oscillator powers for certain relative intensity noise levels when limiting the RIN induced read-out noise to a factor of 10 and a factor of 2, respectively, below the carrier shot noise.

The photodiode impedance additionally depends on the maximum heterodyne frequency.

This unifies all other parameters between the individual concepts – including capacitance and current noise of the photodiodes – and allows for an easier comparison between them.

The procedure is very easy to understand if you compare the influences of shot noise, RIN, and electronic noise in Figure 2.10 on page page 41.

Table 5: RIN levels and corresponding optimal local oscillator powers for eLISA (2013), SAGA and Classic LISA to stay a factor of 10 or a factor of 2 respectively below shot noise.

Parameter		eLISA (2013)	SAGA	Classic LISA	
Carrier shot noise	$\left\langle \widetilde{x}^{sn}_{r/o} \right\rangle_{carrier} =$	5.05×10^{-12}	6.58×10^{-12}	9.90×10^{-12}	$\frac{m}{\sqrt{Hz}}$
RIN (for sn/10)	$RIN_{sn/10} =$	6.35×10^{-10}	4.42×10^{-10}	4.44×10^{-10}	$\frac{1}{\sqrt{Hz}}$
for LO power	$P^{sn/10}_{local} =$	23.20	48.00	47.60	mW
RIN (for sn/2)	$RIN_{sn/2} =$	1.58×10^{-8}	1.10×10^{-8}	1.11×10^{-8}	$\frac{1}{\sqrt{Hz}}$
for LO power	$P^{sn/2}_{local} =$	0.94	1.94	1.89	mW

Without substantial advances in laser, laser amplifier, and photoreceiver technology, electronic noise and relative intensity noise will always be quite close to the observatory's shot noise and are potentially limiting noise sources. It is obvious from the measurements presented in Figure 3.11 that the RIN levels required for an order of magnitude lower read-out noise are not achievable with todays technology. When we aim for the factor-of-two below shot noise RIN levels and use the fiber amplifier measurement with the lowest relative intensity noise as a reference, we find different forbidden frequency domains for the individual mission concepts.

For eLISA (2013) we may not use heterodyne frequencies below $5\,\mathrm{MHz}$ to comply with the $1.58 \times 10^{-8}\,\frac{1}{\sqrt{Hz}}$ RIN requirement. SAGA and Classic LISA both must be limited to heterodyne frequencies above $7\,\mathrm{MHz}$ to stay below a RIN of $1.10 \times 10^{-8}\,\frac{1}{\sqrt{Hz}}$.

3.3 BEAT NOTE FREQUENCY MANAGEMENT

Beat notes of the inter-spacecraft interferometer naturally drift due to the time-varying Doppler shifts caused by a relative velocity in the line-of-sight. Figure 3.12 illustrates as an example for eLISA (2013) and Classic LISA that— even for specially chosen and highly optimized orbits [108]—this effect causes variations of up to 2 % of the nominal arm length. The large minimum heterodyne frequencies derived in the previous section turn out to be quite challenging since we have to manage all beat notes on board each spacecraft at once. The goal now is not only to avoid zero heterodyne frequencies but additionally to keep all frequencies above the 5 MHz or even 7 MHz level, while at the same time restraining the maximum heterodyne frequency to a reasonable range.

To keep the drifting beat notes under control, adaptable offset frequency phase-locked loops between the different lasers will be implemented. Every laser is frequency locked to another one, with only one laser (the "master") being locked to a stable reference. All of these locks have an adaptable offset frequency. As a result, many beat notes can be kept at a constant frequency since even Doppler shifts are compensated by the locking control loop. Only a few remaining beat notes are subject to frequency drifts that cannot be controlled directly.

The **locking control loop** cannot distinguish between frequency noise of the laser it locks to and frequency drifts introduced by Doppler shifts.

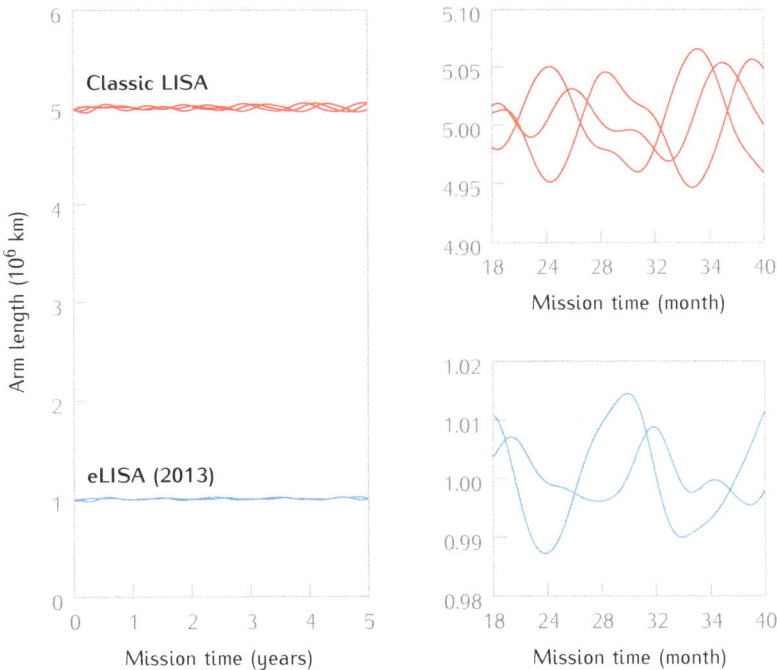

Figure 3.12: Variation of inter spacecraft separation distance due to disturbances within our solar system for Classic LISA and eLISA. All variations can be kept within 2 % of the nominal arm length. Superimposed on these large (but very slow) distance variations would be much smaller variations of higher frequency due to gravitational waves of a multitude of distant sources. The impact of these gravitational waves within a frequency range of roughly 0.10 mHz to 1 Hz is on the order of sub-nm for most sources.

3.3.1 LOCKING SCHEMES

For a three-arm observatory, three different carrier beat notes exist at each spacecraft. Which beat note is constant and which is Doppler shifted over time depends on the particular locking scheme.

EXAMPLE: One possible locking scheme is illustrated in Figure 3.13. Here Laser (A) on board spacecraft S/C 1 is chosen to be the master laser. Laser (B) on board the same spacecraft is locked to Laser (A). Hence the beat note between (A) and (B) at S/C 1 can be kept constant at offset frequency $f_{AB,1}(t) = \Delta f_{11}$. Laser (F) on board spacecraft S/C 3 is locked to the incoming light of Laser (A). Hence Doppler shifts $f_{d3}(t)$ between S/C 1 and S/C 3 are compensated by the locking control loop and the beat note between (A) and (F) at S/C 3 is again constant at offset frequency $f_{AF,3}(t) = \Delta f_{31}$. The beat note between the very same lasers at S/C 1, $f_{AF,1}(t)$, though is even more subject to Doppler shifts: due to the frequency lock of (F) to the Doppler shifted light received from (A), the frequency of (F) now intrinsically shifts with $f_{d3}(t)$. If sent back to S/C 1, the influence of the Doppler shift doubles which results in $f_{AF,1}(t) = \Delta f_{31} + 2 \times f_{d3}(t)$. Uncontrolled beat notes between lasers further away from the master laser in the locking scheme can have a more complex combination of many offset frequencies and different Doppler shifts.

There are only four different uncontrolled beat notes in the entire constellation, but many different locking schemes are possible. Three exemplary schemes with different dependencies are illustrated in Figure 3.14, where the bottom scheme corresponds to the one shown in Figure 3.13. Additionally the locking direction and the position of the master laser can be altered.

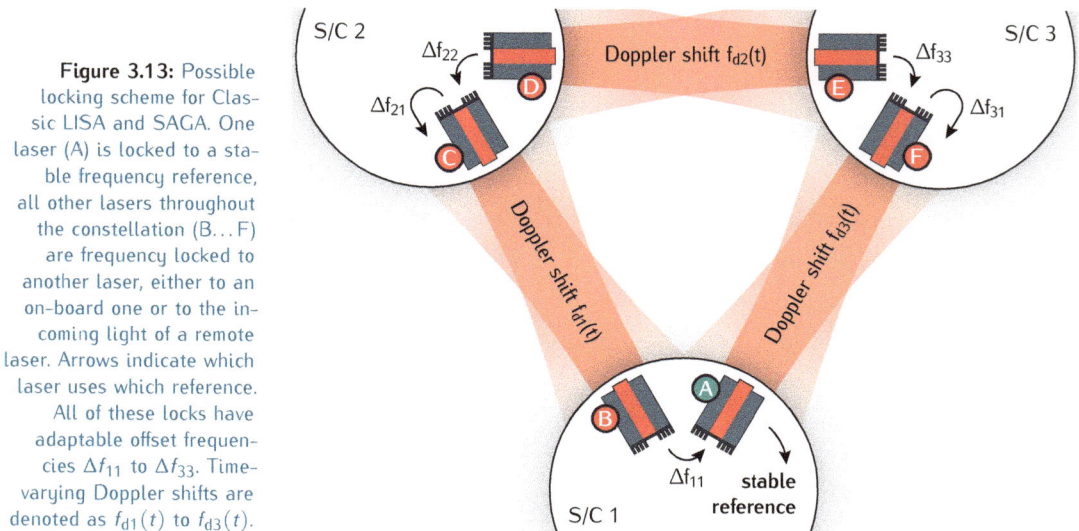

Figure 3.13: Possible locking scheme for Classic LISA and SAGA. One laser (A) is locked to a stable frequency reference, all other lasers throughout the constellation (B...F) are frequency locked to another laser, either to an on-board one or to the incoming light of a remote laser. Arrows indicate which laser uses which reference. All of these locks have adaptable offset frequencies Δf_{11} to Δf_{33}. Time-varying Doppler shifts are denoted as $f_{d1}(t)$ to $f_{d3}(t)$.

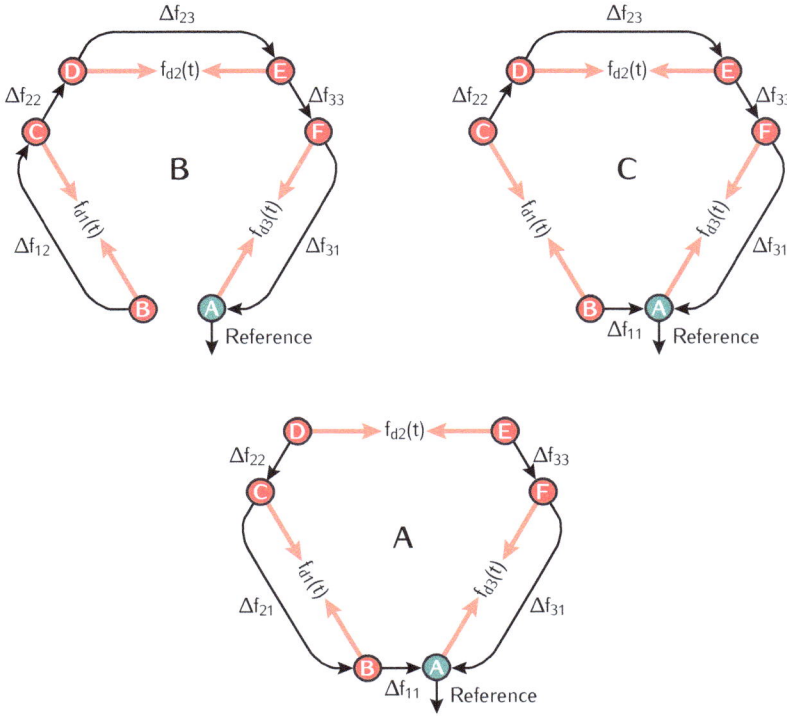

Figure 3.14: Three different locking scheme for Classic LISA and SAGA. One laser (A) is locked to a stable frequency reference, all other lasers throughout the constellation (B...F) are frequency locked to another laser, either to an on-board one or to the incoming light of a remote laser. Arrows indicate which laser uses which reference. All of these locks have adaptable offset frequencies Δf_{11} to Δf_{33}. Time-varying Doppler shifts are denoted $f_{d1}(t)$ to $f_{d3}(t)$. The bottom scheme corresponds to the one shown in Figure 3.13.

As described in the earlier example, we can state the beat note frequencies for all nine beat notes in the constellation as a function of the time-varying Doppler shifts $f_{d1}(t) \dots f_{d3}(t)$ and offset frequencies $\Delta f_{11} \dots \Delta f_{33}$. This was done for all three locking schemes.

Locking scheme A:

$$
\begin{aligned}
f_{\mathrm{AB},1}(t) &= \Delta f_{11} \\
f_{\mathrm{AF},3}(t) &= \Delta f_{31} \\
f_{\mathrm{AF},1}(t) &= \Delta f_{31} + 2 \times f_{d3}(t) \\
f_{\mathrm{BC},2}(t) &= \Delta f_{21} \\
f_{\mathrm{BC},1}(t) &= \Delta f_{21} + 2 \times f_{d1}(t) \\
f_{\mathrm{EF},3}(t) &= \Delta f_{33} \\
f_{\mathrm{CD},2}(t) &= \Delta f_{22} \\
f_{\mathrm{DE},3}(t) &= f_{d3}(t) + \Delta f_{31} + \Delta f_{33} \\
&\quad - [\Delta f_{11} + f_{d1}(t) + \Delta f_{21} + \Delta f_{22} + f_{d2}(t)] \\
f_{\mathrm{DE},2}(t) &= f_{d3}(t) + \Delta f_{31} + \Delta f_{33} + f_{d2}(t) \\
&\quad - \left[\Delta f_{11} + f_{d1}(t) + \Delta f_{21} + \Delta f_{22} \right]
\end{aligned}
\tag{65}
$$

Locking scheme B:

$$
\begin{aligned}
f_{AB,1}(t) &= f_{d1}(t) + \Delta f_{12} + \Delta f_{22} + f_{d2}(t) + \Delta f_{23} \\
&\quad + \Delta f_{33} + f_{d3}(t) + \Delta f_{31} \\
f_{AF,3}(t) &= \Delta f_{31} \\
f_{AF,1}(t) &= \Delta f_{31} + 2 \times f_{d3}(t) \\
f_{BC,2}(t) &= \Delta f_{12} + 2 \times f_{d1}(t) \\
f_{BC,1}(t) &= \Delta f_{12} \\
f_{EF,3}(t) &= \Delta f_{33} \\
f_{CD,2}(t) &= \Delta f_{22} \\
f_{DE,3}(t) &= \Delta f_{23} + 2 \times f_{d2}(t) \\
f_{DE,2}(t) &= \Delta f_{23}
\end{aligned}
$$

(66)

Locking scheme C:

$$
\begin{aligned}
f_{AB,1}(t) &= \Delta f_{11} \\
f_{AF,3}(t) &= \Delta f_{31} \\
f_{AF,1}(t) &= \Delta f_{31} + 2 \times f_{d3}(t) \\
f_{BC,2}(t) &= \Delta f_{11} + f_{d1}(t) - \Big[\Delta f_{31} + f_{d3}(t) \\
&\quad + \Delta f_{33} + \Delta f_{23} + f_{d2}(t) + \Delta f_{22} \Big] \\
f_{BC,1}(t) &= \Delta f_{11} - \Big[\Delta f_{31} + f_{d3}(t) + \Delta f_{33} \\
&\quad + \Delta f_{23} + f_{d2}(t) + \Delta f_{22} - f_{d1}(t) \Big] \\
f_{EF,3}(t) &= \Delta f_{33} \\
f_{CD,2}(t) &= \Delta f_{22} \\
f_{DE,3}(t) &= \Delta f_{23} + 2 \times f_{d2}(t) \\
f_{DE,2}(t) &= \Delta f_{23}
\end{aligned}
$$

(67)

In the following we will refer to the set of beat notes of Equation 65 (Figure 3.14, bottom) as 'locking scheme A'. Beat notes of Equation 66 (Figure 3.14, top left) and Equation 67 (Figure 3.14, top right) are referred to as 'locking scheme B' and 'locking scheme C', respectively.

Actually, a second local laser is necessary in the daughter spacecraft as well so that basically **two carrier beat note frequencies** exist. The second beat note can be set to an arbitrary and constant value though.

A two arm observatory makes the situation a little simpler. Here, only the mother spacecraft has to deal with three carrier beat note frequencies. In the two daughter spacecraft only one beat note exists. Two possible locking schemes for a two arm observatory are illustrated in Figure 3.15. In total, only five beat notes have to be managed as a function of the time-varying Doppler shifts $f_{d1}(t)$ and $f_{d3}(t)$ and offset frequencies $\Delta f_{11} \ldots \Delta f_{31}$. For the two locking schemes, we can state the beat notes as follows.

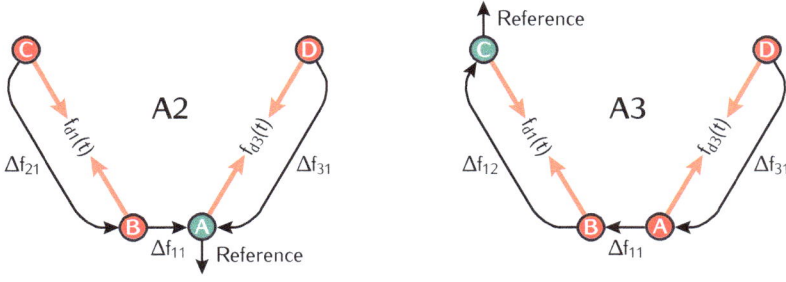

Figure 3.15: Two different locking scheme for eLISA (2013). One laser is locked to a stable frequency reference, all other lasers throughout the constellation are frequency locked to another laser. Arrows indicate which laser uses which reference.

Locking scheme A2:

$$f_{AB,1}(t) = \Delta f_{11}$$
$$f_{AD,3}(t) = \Delta f_{31}$$
$$f_{AD,1}(t) = \Delta f_{31} + 2 \times f_{d3}(t) \qquad (68)$$
$$f_{BC,2}(t) = \Delta f_{21}$$
$$f_{BC,1}(t) = \Delta f_{21} + 2 \times f_{d1}(t)$$

Locking scheme A3:

$$f_{AB,1}(t) = \Delta f_{11}$$
$$f_{AD,3}(t) = \Delta f_{31}$$
$$f_{AD,1}(t) = \Delta f_{31} + 2 \times f_{d3}(t) \qquad (69)$$
$$f_{BC,2}(t) = \Delta f_{12} + 2 \times f_{d1}(t)$$
$$f_{BC,1}(t) = \Delta f_{12}$$

Obviously, the remaining beat notes of Equation 68 (Figure 3.15, left) and Equation 69 (Figure 3.15, right) are quasi identical to locking scheme A and will be referred to as 'locking scheme A2' and 'locking scheme A3', respectively.

3.3.2 DOPPLER SHIFTS

Orbit simulations performed by Oliver Jennrich [108] yield highly optimized parameters to keep the relative velocity in the line-of-sight at a minimum. Simulated data for different arm lengths and mission durations exists. To get a broader picture that exceeds the three considered mission concepts, I used the line-of-sight velocities for arm lengths of 1, 2, 3, and 5×10^6 km for a timespan of 2 and 5 years. All orbits were optimized to keep all three line-of-sight velocities at a minimum. For the case of a two-arm observatory, one of the three parameters was simply ignored. There might be additional room for further optimization taking the now known locking schemes and resulting beat notes into consideration. Yet these orbits are a good basis to test the feasibility of certain locking schemes and address the issue of the maximum heterodyne frequency.

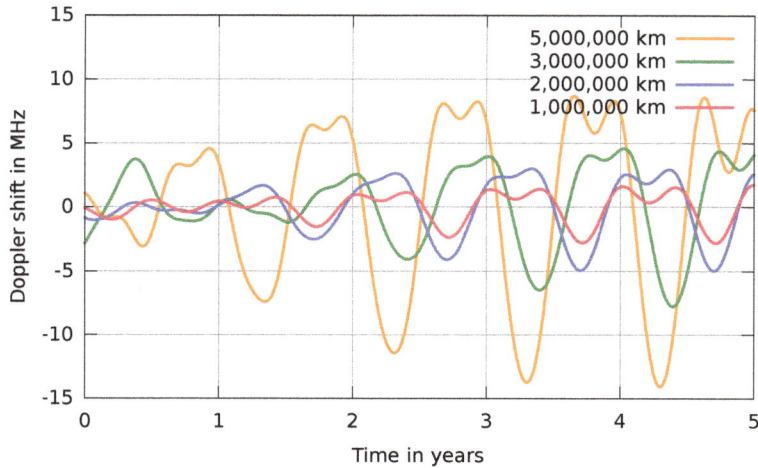

Figure 3.16: Doppler frequency shifts for different arm length over a time of 5 years. Only one Doppler shift per arm length is illustrated.

Simulated line-of-sight velocity data converted to Doppler frequency shifts for a laser wavelength of $\lambda_{\text{laser}} = 1064$ nm (see Equation 8) is plotted in Figure 3.16 over a time of 5 years. Only one Doppler shift per arm length is illustrated.

3.3.3 FREQUENCY PLAN

All offset frequencies need to be switched regularly to keep all beat notes within a certain range, given by the minimum frequency discussed above and an upper limit that should be as low as possible to relax a wide range of system requirements. These switches result in a loss of the laser lock and render the observatory non-operational for a short while. We need to assess the number of necessary frequency switches within the mission duration and the exact offset frequencies for given Doppler shifts at a chosen frequency range. This has never been evaluated before in detail.

I chose a **genetic algorithm** since I never used one before and always wanted to. I am aware of the fact that other algorithms exist that may be more efficient [109]. Computationally, this problem is no match for modern computers either way.

For the case of a two-arm observatory, only **three-gened individuals** are evaluated.

I developed a **genetic algorithm** (source code available in Section A.1 on page 159) that finds a suitable schedule of offset frequencies to keep all beat notes as long as possible within a certain frequency range without a switch in frequencies. A secondary figure of merit is the distance to the bounds of the frequency range.

The algorithm gives birth to an initial population of random but viable (fit for survival for at least one day) individuals. Each individual possesses five "genes". Every gene holds one offset frequency in hertz represented as binary number. The individuals are evaluated for viability (number of days all beat notes can be kept within certain frequency limits at the given set of offset frequencies) and distance to the bounds. The fitter part of the population is allowed to procreate. From this group, all individuals—randomly chosen in pairs—mix genes. At a certain probability, genes of the same type are recombined, otherwise one gene is considered dominant and used as a whole for the offspring. Occasionally, genes mutate – represented by a bit flip within a certain range of the less significant bits.

The functionality of the algorithm can be verified graphically. In Figure 3.17 individuals are plotted in five dimensions, each representing the offset frequency value of one gene in MHz. The example shows results for a three-arm observatory in locking scheme A with frequency bounds of $7\ldots23$ MHz. The viability of each individual is color coded (fitter individuals are red). The algorithm followed 30 generations for each set of offset frequencies. Displayed are the initial population and generations 8, 16, and 30 for day 55. The optimal set of offset frequencies was found in generation 30 and lasts for 103 days without a switch. The result of the entire run is summarized in Table 6. Within the frequency bounds and for a five year mission, only 24 frequency switches are necessary. The shortest duration without a switch is 12 days, the longest undisturbed measurement run would be 177 days if no other events cause a short time failure of the laser lock.

Day	Δf_{11}	Δf_{31}	Δf_{21}	Δf_{33}	Δf_{22}	Days
1	18.59	18.09	−7.00	−18.61	−11.40	15
16	15.84	17.22	9.17	−15.82	−14.68	39
55	14.57	8.51	−11.68	−14.33	−19.01	103
158	−16.38	−16.27	−15.07	−15.73	16.71	16
174	17.69	13.19	−9.35	−18.31	−10.16	25
199	−18.75	−13.20	19.79	19.01	10.44	12
211	−15.24	−14.37	−10.32	−16.06	−15.26	59
270	−16.06	18.93	12.45	−14.94	−15.98	57
327	−15.22	19.34	9.66	14.41	14.86	85
412	−18.68	8.68	−11.71	−18.35	10.49	177
589	15.73	17.17	−13.11	−15.37	−16.40	57
646	−14.41	19.37	−18.49	−11.80	18.35	136
782	−15.08	9.83	17.72	19.58	18.35	141
923	15.73	16.39	−11.13	−15.77	14.82	63
986	−16.40	−10.48	10.47	11.80	−13.10	176
1162	11.79	11.13	−10.79	−13.76	19.01	171
1333	15.87	−10.48	13.48	15.77	−12.44	124
1457	10.95	15.56	8.19	−10.46	−18.32	22
1479	−15.44	−15.06	10.15	−14.90	−14.35	70
1549	−15.81	11.13	18.05	−15.32	17.04	51
1600	−15.74	10.48	20.67	14.01	15.00	87
1687	16.24	−9.17	−17.07	14.92	17.17	117
1804	15.17	−15.73	12.45	15.04	−15.04	18
1822	−18.44	−14.24	10.09	−11.81	−18.48	>5

Table 6: Optimized offset frequencies for an exemplary run of the genetic algorithm over a 5 year mission lifetime with frequency bounds of $7\ldots23$ MHz for a three-arm observatory in locking scheme A. Frequencies are given in MHz. The shortest duration without a switch is 12 days while the longest undisturbed measurement can be performed for 177 days.

The results for a different run with the same parameters is shown in Figure 3.18, including the remaining four beat notes influenced by Doppler shifts. Although for this run the individual offset frequencies are quite different, the main findings (number of switches and days of the frequency switch) are identical. This is a good test for the reproducibility of results generated by the algorithm.

Generation 0

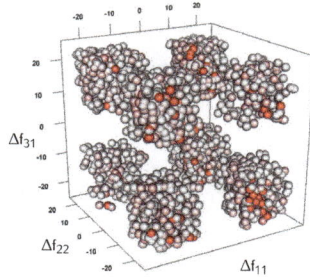

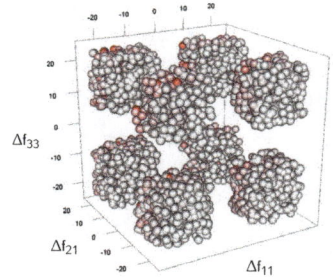

Figure 3.17: Graphical representation of the genetic algorithm used to optimize the frequency plan. This example shows results for a three-arm observatory in locking scheme A with frequency bounds of $7\dots23\,\text{MHz}$. Each dimension represents the offset frequency value of one gene in MHz. Balls are single individuals. The viability (number of days all beat notes can be kept within the given frequency limits for the genetically encoded set of offset frequencies) is color coded (fitter individuals are red). The algorithm followed 30 generations for each set of offset frequencies. Displayed are the initial population and generations 8, 16, and 30 for day 55. The optimal set of offset frequencies was found in generation 30 and lasts for 103 days without a switch. The chosen parameters were: 4000 individuals per generation, 33% allowed to procreate, 40% gene recombination probability, 5% gene mutation probability within the 16 least significant bits.

simonbarke.com/phd/gen

Generation 8

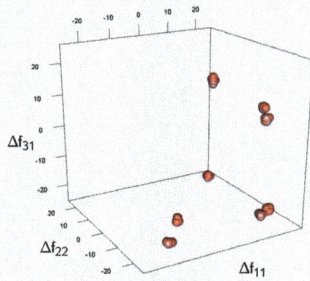

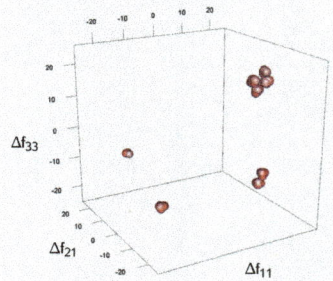

Generation 16

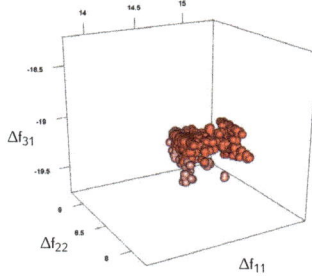

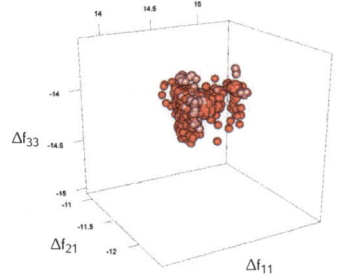

Generation 30

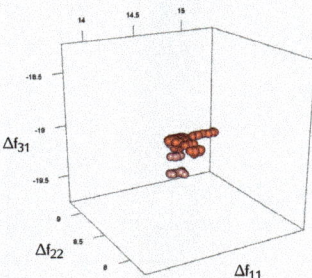

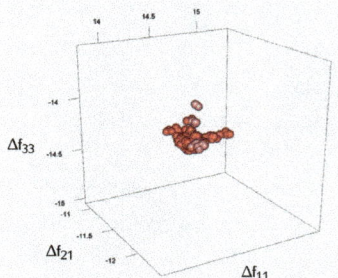

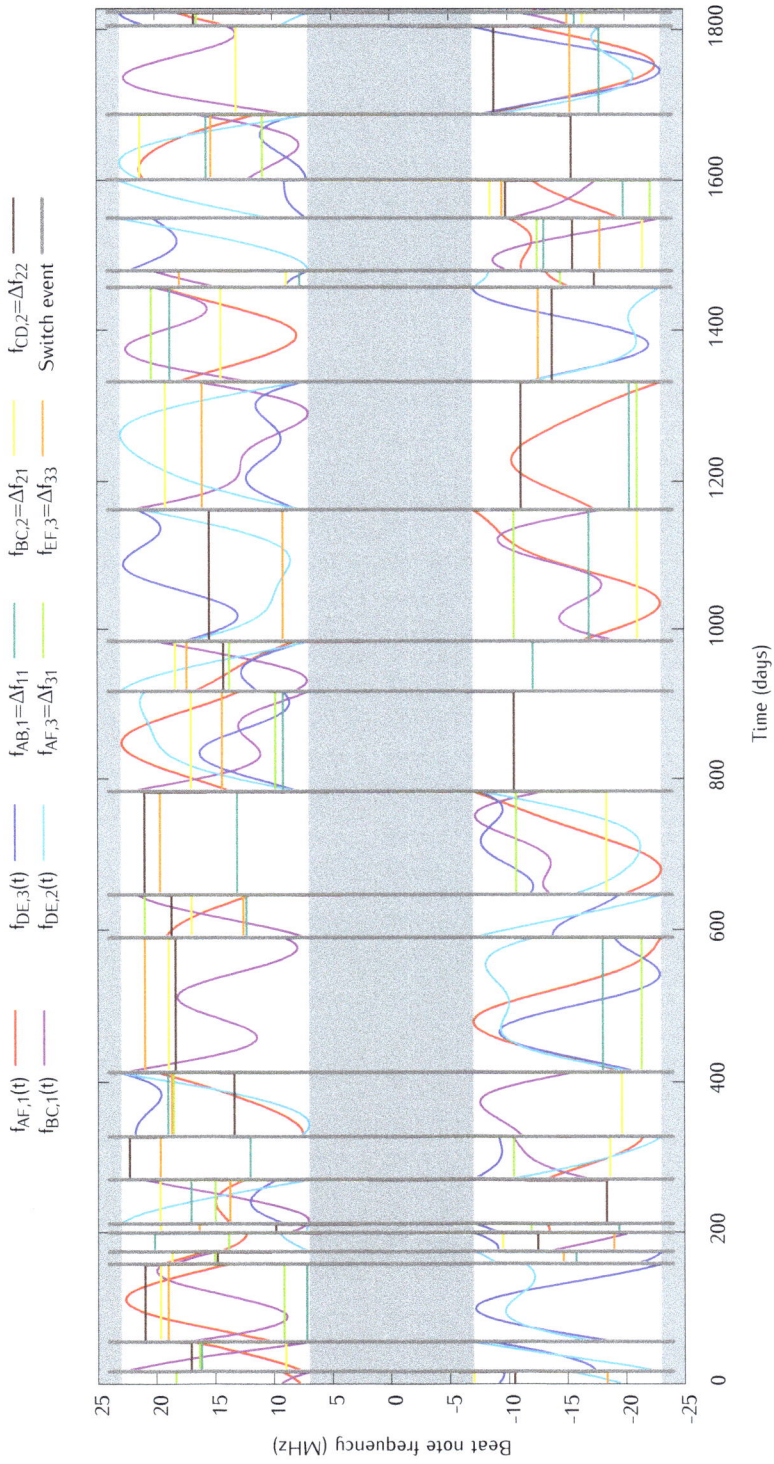

Figure 3.18: Optimized offset frequencies for an exemplary run of the genetic algorithm over a 5 year mission lifetime with frequency bounds of 7 ... 23 MHz for a three-arm observatory in locking scheme A. Horizonal line segments represent programmed fixed frequency offsets in a phase locked loop. The remaining traces are beat notes influenced by Doppler shifts. The shortest duration without a switch is 12 days, the longest undisturbed measurement can be performed over 177 days.

Different parameters were tested graphically to ensure that local clusters of fit individuals do not die out prematurely, unfit offspring does not procreate uncontrolled, and optimal offset frequencies are found efficiently. The example uses 4000 individuals per generation of which 33% were allowed to procreate with a 40% gene recombination probability and a 5% gene mutation probability within the 16 least significant bits. Each offset frequencies was represented by a 32 bit value.

In the end, a different set of parameters turned out to yield much faster yet equally reliable results with a population of only 200 individuals over 20 generations. The 50 fittest percent of each generation were allowed to procreate with a gene recombination probability of 40%. Mutation parameters remained unchanged.

3.3.4 MAXIMUM HETERODYNE FREQUENCY

The main purpose of the developed algorithm was not the creation of a frequency plan. The relative line-of-sight velocity data is not yet optimized for the different locking schemes. Additionally, frequency switches should coincide with other system maintenance events such as radio antenna repointings. Instead, we can use the algorithm to find a lower bound for the maximal beat note frequency. This value is an issue of universal concern since it influences many different requirements.

I did 108 runs for different beat note frequency ranges using a lower bound of $5 \ldots 13\,\text{MHz}$ and an upper bound of $19 \ldots 30\,\text{MHz}$. All runs were performed with simulated Doppler frequency data for $5\,000\,000\,\text{km}$ arm length, all three arms were considered over a total duration of 5 years in locking scheme A. The results are visualized as matrix in Figure 3.19. Green stands

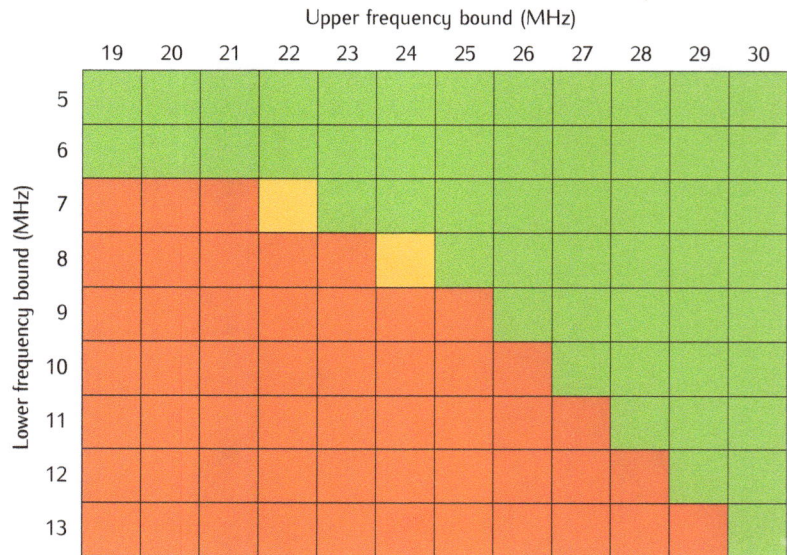

Figure 3.19: Matrix of 108 algorithm runs for different beat note frequency ranges (5 000 000 km arm length, three arms, 5 years, locking scheme A). Green stands for a frequency plan of similar quality to the one presented in Table 6 where frequencies are switched every few weeks. Yellow implies switching on an almost daily basis. Red means that no set of offset frequencies could be found to keep all beat notes within the given bounds.

for a frequency plan of similar quality to the one presented in Table 6 where frequencies are switched every few weeks. Yellow implies switching on an almost daily basis. Red means that no set of offset frequencies could be found to keep all beat notes within the given bounds. The higher the lower bound, the harder it becomes to find viable solutions.

Results for other locking schemes were very similar so that the choice of a certain locking scheme can be based on other motivations. Since locking scheme A and A2 are symmetrical and—starting from the master laser—can reinstate the lock for one half of the constellation before dropping the other half, these schemes are preferred and will be used solely from now on.

Let's consider the determined forbidden frequency domains for eLISA (2013), SAGA, and Classic LISA with minimum frequencies of 5 and 7 MHz in a more general scope. Table 7 shows the smallest possible maximum heterodyne frequencies for a five year mission of 1, 2, 3, and 5×10^6 km observatories with 2 and 3 arms. For comparison the best achievable values are given for an eLISA-like mission with a nominal lifetime of 2 years. To be on the safe side – and to avoid the frequency range dominated by relative intensity noise and allow for sideband beat notes and other auxiliary functions – I specified lower frequency bounds of 7 and 9 MHz to the algorithm. Keep in mind that there will be additional sideband beat notes 1 MHz above and below the main heterodyne frequency as described in Section 4.3.2. Thus we have to add a buffer of at least 1 MHz to the upper and lower frequency bounds for all other purposes. This buffer was already added in Table 7.

As we will see later in Section 4.3.1 it is beneficial to add **another tone** at exactly 5 MHz that must not overlap with any carrier or sideband beat notes to allow for noise suppression [65].

Arm length		Five-year Lower bound:	Two-arm 6 MHz	8 MHz	Three-arm 6 MHz	8 MHz
1×10^6 km		Upper bound:	13 MHz	15 MHz	14 MHz	16 MHz
2×10^6 km			16 MHz	18 MHz	18 MHz	20 MHz
3×10^6 km			18 MHz	20 MHz	23 MHz	26 MHz
5×10^6 km			22 MHz	25 MHz	24 MHz	28 MHz
		Two-year				
1×10^6 km			10 MHz	12 MHz	12 MHz	14 MHz

Table 7: Smallest possible maximum heterodyne frequencies (upper bounds) for two different lower bounds (6 and 8 MHz). 2 and 5 year missions (1, 2, 3, and 5×10^6 km arm length), two- and three-arm observatories.

The upper bounds vary greatly for all combinations, yet they are throughout within reasonable limits. Current technology for the signal read-out can be adapted to measure the phase of beat notes of up to 30 MHz [74].

3.4 MISSION CONCEPT REQUIREMENTS

So far, in this chapter, we have acquired detailed knowledge about

✦ the carrier signal read-out noise level including shot noise for all three observatory concepts (Table 4)

✦ the assumed relative intensity noise levels for a RIN-induced read-out noise, a factor of two below shot noise (Table 5), and

✦ the associated upper and lower beat note frequency bounds (Table 7).

With this information, we can now not only state the equivalent displacement noise from Equation 14 but additionally calculate an equivalent timing noise from Equation 35 using the corresponding maximum heterodyne frequency. This is done in Table 8 for all three mission concepts.

Table 8: Carrier shot noise given as phase noise, displacement noise, and timing noise equivalents alongside relative intensity noise and heterodyne frequency bounds for different mission concepts.

Noise relevant parameters		eLISA (2013)	SAGA	Classic LISA	
Shot noise	$\left\langle \tilde{x}^{sn}_{r/o} \right\rangle_{carrier} =$	5.05×10^{-12}	6.58×10^{-12}	9.90×10^{-12}	$\frac{m}{\sqrt{Hz}}$
RIN	$RIN =$	1.58×10^{-8}	1.10×10^{-8}	1.11×10^{-8}	$\frac{1}{\sqrt{Hz}}$
Lower freq.	$f_{min} =$	6	8	8	MHz
Upper freq.	$f_{het} =$	13	20	28	MHz
Phase noise	$\left\langle \tilde{\phi}^{sn}_{r/o} \right\rangle_{carrier} =$	1.49×10^{-5}	1.94×10^{-5}	2.92×10^{-5}	$\frac{rad}{\sqrt{Hz}}$
Timing noise	$\left\langle \tilde{t}^{sn}_{r/o} \right\rangle_{carrier} =$	1.82×10^{-13}	1.54×10^{-13}	1.66×10^{-13}	$\frac{s}{\sqrt{Hz}}$

These noise levels lead to requirements that are of vital significance for the final part of this thesis. Similar to the relative intensity noise we have to set limits for the allowed excess and residual noise introduced by the many different systems, components, and data processing tools. As a rule of thumb, it should always be desirable to keep the influence of individual uncorrelated noise sources one order of magnitude below the total carrier signal read-out phase noise. Updated values for this noise can be found in Table 9 alongside ambitious requirements that are a factor of ten below the corresponding value.

The **total** read-out noise includes influences of RIN and photoreceiver electronics.

Table 9: System requirements (factor ten below total read-out noise) for phase noise, equivalent displacement noise, and timing noise. Values are valid above a particular corner frequency as indicated.

Requirements		eLISA (2013)	SAGA	Classic LISA	
Read-out noise	$\left\langle \tilde{x}^{total}_{r/o} \right\rangle =$	6.21×10^{-12}	7.90×10^{-12}	1.24×10^{-11}	$\frac{m}{\sqrt{Hz}}$
Dominates above	$f_{corner} =$	4.95×10^{-3}	4.39×10^{-3}	3.50×10^{-3}	Hz
Phase Req.	$\left\langle \tilde{\phi}^{/10}_{req} \right\rangle =$	3.67×10^{-6}	4.67×10^{-6}	7.32×10^{-6}	$\frac{rad}{\sqrt{Hz}}$
Displacement Req.	$\left\langle \tilde{x}^{/10}_{req} \right\rangle =$	6.21×10^{-13}	7.90×10^{-13}	1.24×10^{-12}	$\frac{m}{\sqrt{Hz}}$
Timing Req.	$\left\langle \tilde{t}^{/10}_{req} \right\rangle =$	4.49×10^{-14}	3.71×10^{-14}	4.16×10^{-14}	$\frac{s}{\sqrt{Hz}}$

All values are based on todays photoreceiver and laser technology. Advances in both fields would lower the actual relative intensity noise and reduce the lower and upper frequency bounds. This would increase the equivalent timing noise and **allow for relaxed requirements**.

The phase fidelity requirements (and its equivalent displacement noise value) differ between the individual mission concepts by almost a factor of two. The associated timing stability requirement for signals at the maximum heterodyne frequency turns out to be almost identical for all missions.

All requirements are valid only for Fourier frequencies where the readout noise level is higher than the expected acceleration noise [110] which increases proportionally to f^{-2} towards lower frequencies and hence quickly becomes the dominating influence (see Figure 2.18). We should keep in mind that the acceleration noise scales differently in the full observatory sensitivity. It accounts for a noise contribution two times higher than all other displacement noise sources (see Equation 61). Hence, to find the frequencies at which acceleration takes over as dominating influence, I used a level of $2 \times \tilde{x}_{\mathrm{acc}}(f) = 6 \times 10^{-15} \frac{\mathrm{m/s^2}}{\sqrt{\mathrm{Hz}}} \times \frac{1}{(2\pi f)^2} \stackrel{!}{=} \left\langle \tilde{x}_{\mathrm{req}}^{\prime 10} \right\rangle$ and solved for f. These corner frequencies f_{corner} are summarized in Table 9 as well. The final requirements (valid in the targeted gravitational wave frequency range of roughly $f = 10^{-4} \ldots 1\,\mathrm{Hz}$) then come down to

$$\left\langle \tilde{\phi}_{\mathrm{req}}^{\prime 10} \right\rangle (f) = \left\langle \tilde{\phi}_{\mathrm{req}}^{\prime 10} \right\rangle \times \sqrt{1 + \left(\frac{f_{\mathrm{corner}}}{f} \right)^4}. \tag{70}$$

Frequency dependent equivalent displacement noise and timing stability requirements can be constructed similarly. The real level of acceleration noise will become much clearer after a successful LISA Pathfinder mission. For now, this approximation will be good enough for the evaluation of different components.

Figure 3.20 comprises the phase noise requirements and their displacement noise equivalents for all three mission concepts. It becomes visible how smaller and supposedly more cost efficient mission concepts actually are more demanding and put tougher requirements on all systems. In up-

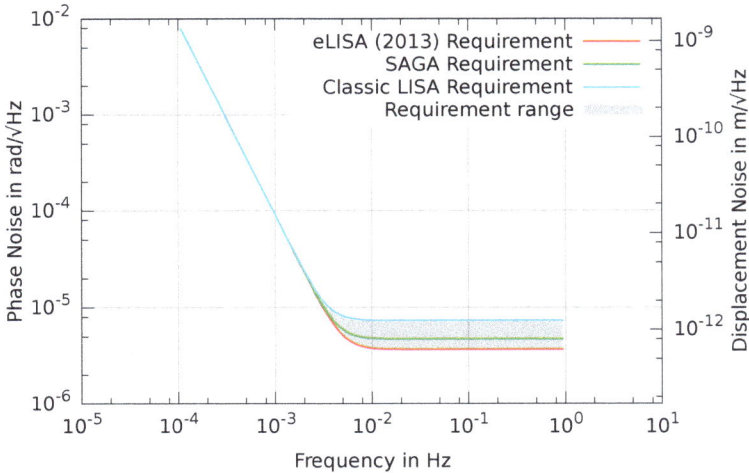

Figure 3.20: Phase noise requirements and their displacement noise equivalents for all three mission concepts. The gray area indicates how phase and displacement noise requirements will be indicated in measurement results throughout this thesis.

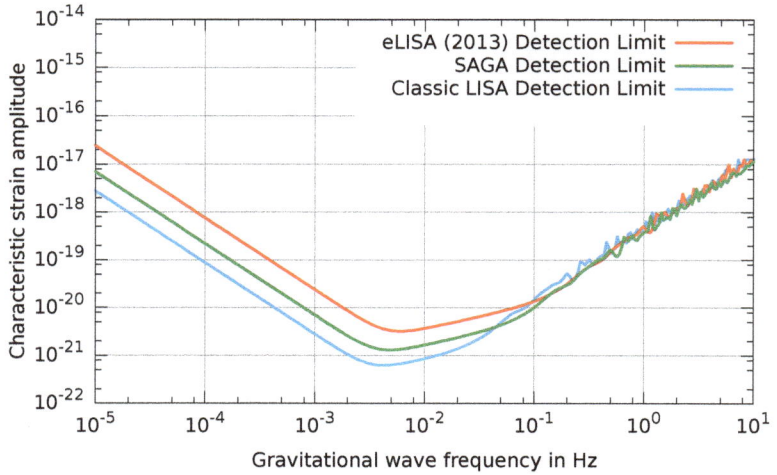

Figure 3.21: Detection limits for eLISA (2013), SAGA, and Classic LISA (SNR=1) given in characteristic strain amplitude.

The **detection limits** were determined by the Gravitational Wave Observatory Designer (see Section 2.3.3) with values from Table 8.

coming measurements the many different requirements would distract from the actual results. Thus I will use a requirement range similar to the gray area throughout this thesis. The resulting detection limits (where the signal-to-noise ratio equals 1) given in characteristic strain amplitude is shown in Figure 3.21. Naturally, Classic LISA offers the best sensitivity, followed by SAGA and eLISA (2013).

The requirements are converted to timing noise in Figure 3.22. Here, requirements for eLISA (2013) are relaxed the most while SAGA and Classic LISA are quite similar. Again, a gray requirement range as indicated will be used from now on.

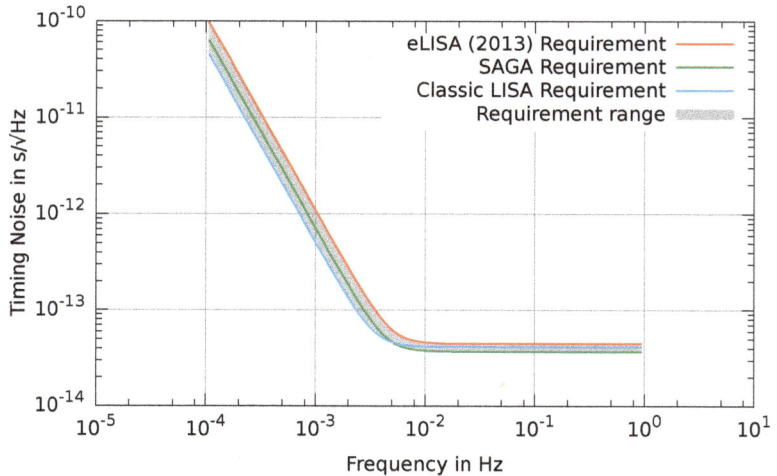

Figure 3.22: Timing stability requirements for all three mission concepts. The gray area indicates how these requirements will be indicated in measurement results throughout this thesis.

The knowledge about the timing stability requirements and their dependence on the heterodyne frequency range can be used as additional optimization criteria for mission concept studies like the one performed by *Airbus Defence and Space* which resulted in the parameter set the SAGA concept was based on. In general, it might be worthwhile to come up with a mission concept that keeps the timing stability requirements as relaxed as possible which in turn will ease the challenges of technology development and testing. This can be done by reducing the maximum heterodyne frequency (which means shrinking the arm length) or by lowering the received light power (which will relax the phase noise requirements the timing stability is based on). Both measures potentially lower the observatory's detection limit and a careful balance of optical power, telescope diameter, and arm length within technical and financial constraints is necessary.

For example, one could think of a 3-arm observatory (let's name it 'SAGA+') with $L_{arm} = 2\,500\,000$ km arm length and an optical power of $P_{tel} = 2$ W sent through telescopes $d_{tel} = 23$ cm in diameter. The sensitivity would be on par with SAGA although the equivalent displacement noise requirements are relaxed to $\left\langle \widetilde{x}_{req}^{\prime 10} \right\rangle = 1.17 \times 10^{-12}$ m/$\sqrt{\text{Hz}}$ for a relative intensity noise of $1.14 \times 10^{-8}/\sqrt{\text{Hz}}$ which is a factor of 2 below the carrier read-out noise. This is due to the fact that the received light power is significantly lower, in fact very similar to the original Classic LISA concept. As a result, the SAGA+ observatory would feature an easiest-in-class timing requirement of $\left\langle \widetilde{t}_{req}^{\prime 10} \right\rangle = 4.78 \times 10^{-14}$ s/$\sqrt{\text{Hz}}$ for a maximum heterodyne frequency of estimated $f_{het} = 23$ MHz.

Similar to eLISA (2013), the smaller telescopes reduce the total mission cost due to an overall more compact—lighter—design. However, the laser power of > 2 W required for the SAGA+ mission might be challenging due to the increased phase noise level of high power optical fiber amplifiers (see Section 5.1.4) and the limited electrical power due to the smaller solar panels of the overall more compact spacecraft dimensions. The very same challenges apply to the eLISA (2013) mission concept.

No matter the final mission design, we can be quite confident that in the end the phase noise and timing stability requirements will fall somewhere within the depicted ranges. Thus all components that were evaluated are compared to the full requirement range for both, phase noise and timing stability requirements. The latter are very handy when dealing with the fidelity of signals from auxiliary functions. Good examples are on-board reference oscillators that run at frequencies considerably higher than the maximum heterodyne frequency and all components that deal with signals in the pilot tone generation and transmission chain. It is related to the Inter-Spacecraft Frequency Distribution System and uses a wide range of frequencies all the way up to several GHz. Phase noise requirements for any of these frequencies relax by the ratio of the actual signal frequency over the maximum heterodyne frequency. This usually complicates the performance comparison

Gravitational Wave Observatory Designer SAGA+ preset with correct heterodyne and modulation frequencies.

spacegravity.org/designer/ #rc=9ef9-8feb-44b6

The SAGA+ detection limit is even **better than SAGA** for low gravitational wave frequencies due the slightly longer arm length.

when considering different mission concepts. In the presented case however, timing jitter does not change with signal frequency since the timing stability requirements already include a scaling factor that arises from the maximum heterodyne frequency. Thus timing stability requirements make it possible to easily compare devices that handle different frequencies.

All of these frequencies are necessary to refer the measurements on board the different spacecraft throughout the constellation to one single constellation-wide reference signal. The need for such a system is dealt with in depth in the next chapter. ∎

Part III

CONSTELLATION-WIDE REFERENCE SIGNAL

Gravitational waves act on the proper distance between the different gravitational reference sensors in the constellation. These proof masses are considered to be the hearts of the spacecraft. If that is true, then the metrology systems clearly are their brains.

All interferometric beat notes for the distance measurements end up here, together with a multitude of signals required for auxiliary functions. Most of these additional signals are used to suppress the influence of otherwise limiting noise sources: There are no oscillators available stable enough so that each spacecraft could use its own reference frequency. On top of that, even within a single spacecraft, analog-to-digital converters add more phase noise to the measurement than allowed by the stringent requirements. To address both issues, each and every measurement has to be augmented with a reference signal (the "pilot tone") that is distributed throughout the entire constellation.

The conversion between reference frequencies and their distribution between all spacecraft to synchronize the different metrology systems is the job of the Inter-Spacecraft Frequency Distribution System. In Chapter 4 I explain the principle of this system in detail and elaborate on the tasks it has to perform. Subsequently, I present results from component evaluation campaigns. In an iterative process we were able to design, develop and successfully test the first fully functional Inter-Spacecraft Frequency Distribution System for gravitational wave observatories. This includes the generation of the pilot tone, required frequency conversions, and the full transmission chain between the local and the remote spacecraft. Additionally, auxiliary functions like the generation of a differential system clock at a frequency different from the pilot tone were developed. All of this functionality is provided by the final system and is subsumed under the heading 'frequency conversion and transmission chain' in Chapter 5.

METROLOGY SYSTEM SYNCHRONIZATION

The metrology system of a gravitational wave observatory has to perform a wide range of tasks, from laser locking and offset frequency control to inter-spacecraft ranging and data transfer. At its core is the Phase Measurement System (or phasemeter) that measures the phase of all digitized interferometric beat notes with microradian precision.

Phase fluctuations can later be processed [91], combined [88], and converted to equivalent changes in the proper distance between any two proof masses. In the end, the effects of gravitational waves will be revealed. Yet this can only work properly if all measurements on the different spacecraft are synchronized to one single reference frequency.

4.1 PHASE MEASUREMENT SYSTEM

Many techniques to measure a signal's phase—like zero-crossing or down-mixing—have been discarded due to significant disadvantages [74]. A digital phase-locked loop (DPLL) operating directly at the signal frequency is currently the preferred architecture for the Phase Measurement System (PMS).

The chosen PLL principle for measuring a signal's phase is shown in Figure 4.1. The input signal is mixed with a sine wave of equal frequency and phase which is produced by an adjustable oscillator. This generates products near DC and at twice the signal frequency f. A low-pass filter removes the $2f$ component and other spurious signals.

For an analog PLL implementation, this **adjustable oscillator** would be, e.g, a voltage controlled oscillator (VCO).

The DPLL uses digital multipliers and filters and a numerical controlled oscillator (NCO) which form one DPLL core as shown in Figure 4.2. Input signals are digitized by an analog-to-digital converter (ADC). Such a DPLL

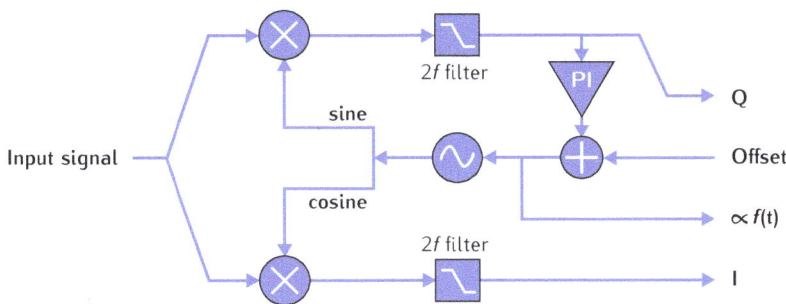

Figure 4.1: In an analog PLL a signal is mixed with a sine and cosine produced by an adjustable oscillator. A PI controller locks onto the input signal and keeps the sine channel at zero by actuating the oscillator's frequency. This information can be used to reconstruct the signal's phase while the cosine channel represents the signal's amplitude.

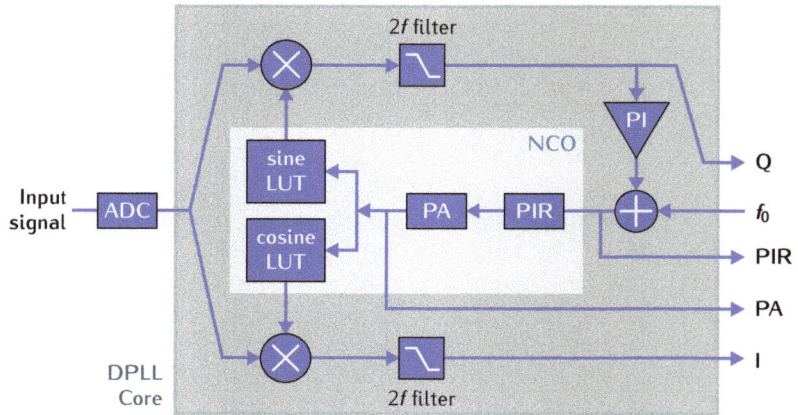

Figure 4.2: In a digital PLL all elements are digital. The oscillator (NCO) is controlled by the value in the PIR and the current value of the PI controller. In combination they set the value in the phase increment register (PIR) which holds the frequency information for phase reconstruction. The phase accumulator (PA) value is used to obtain the instantaneous phase for Differential Wavefront Sensing (DWS).

can track a signal even when its frequency changes due to Doppler shifts. Additionally, the bandwidth can be tuned to allow the independent tracking of several beat notes in a single input signal.

Independent **simultaneous tracking** is made possible by using several DPLL cores on the same digitized signal.

Q MIXER OUTPUT: A proportional-integral (PI) controller keeps the mixer output (Q) very close to zero by actuating the oscillators frequency. Thus this technique tracks the signal's instantaneous frequency $f(t)$. The purpose is to produce a precisely tracking copy of the incoming sine wave, but in digital form.

PIR VALUE: The NCO consists of a phase increment register (PIR), a phase accumulator (PA), as well as a sine and a cosine look up table (LUT). Its instantaneous output frequency is determined by the current value in the PIR and the current value of the PI controller. This frequency data is used to reconstruct the signal's phase changes over time $\phi(t)$ [74, p. 23].

I MIXER OUTPUT: When mixing the input signal with a sine wave 90° out of phase (cosine) the low-pass filtered mixer output (I) represents the instantaneous amplitude of the input signal. This information is beneficial in many ways. For example it can be used to correct for imperfections of the PLL where Q is non-zero but describes the residual difference between the phase of the actual input signal and that of the oscillator.

The **instantaneous amplitude** is assumed to be only slowly varying in normal operations.

PA VALUE: Using the PIR, Q and I values, the signal's phase can be reconstructed only with an unknown offset. The instantaneous phase is, however, available in the phase accumulator (PA) value. It holds very precise spacecraft and proof-mass alignment information. This information can moreover be used for Differential Wavefront Sensing (DWS) which utilizes the PA value to recover the differential phase between signals on quadrant photodiodes.

In combination, the 4 quantities PIR, PA, Q, and I are a complete description and thereby a digital replica of the input signal's sinusoidal component at the frequency of interest.

In order to reach the required fidelity of

$$\left\langle \widetilde{\phi}_{req}^{/10} \right\rangle \left(f = 10^{-4} \ldots 1\,\mathrm{Hz} \right) = \left\langle \widetilde{\phi}_{req}^{/10} \right\rangle \times \sqrt{1 + \left(\frac{f_{corner}}{f} \right)^4}$$

(see Equation 70), **we have to make sure that no component or algorithm in the signal path adds excess phase noise or equivalent timing jitter above this level.**

4.2 AUXILIARY FUNCTIONS

In addition to the primary length measurement, the metrology system has to perform a wide range of auxiliary functions. Among those are beat note acquisition, offset frequency laser locking, and inter-spacecraft ranging and data transfer.

4.2.1 BEAT NOTE ACQUISITION AND LASER LOCKING

The metrology system controls the offset frequencies of the phase-locked loop for all local lasers. Several switches are required to keep all beat notes within a certain frequency range as described in Section 3.3. Additionally the system must be able to perform an automatic lock acquisition. For this purpose, a Fast Fourier Transform (FFT) in combination with a peak finding algorithm is integrated. The frequency of the local laser is adjusted until a carrier beat note is identified. The DPLL can then lock onto this exact frequency.

4.2.2 RANGING

As described in Section 2.2.5, we need to know the absolute distances between any two spacecraft to suppress laser frequency noise [88, 95, 96, 111, 112]. Since the Deep Space Network cannot triangulate the individual spacecraft position accurately enough, an active measurement of the absolute distance between spacecraft based on the principles of the Global Positioning System (GPS) was suggested. The system uses a Direct Sequence Spread Spectrum (DS/SS) modulation. Pseudo random noise (PRN) codes are modulated onto the outgoing laser beams, detected by the receiving spacecraft and compared to a local copy of the same code by the individual metrology systems [75, 90]. The codes were specially designed so that their autocorrelation is close to zero when misaligned and has a sharp maximum when in perfect alignment.

4.2.3 DATA TRANSFER

The ranging mechanism can be enhanced with data encoding. The PRN code is divided into individual data periods onto which the data bits are modulated with an XOR operation. The metrology system on the receiving spacecraft then decodes the data upon arrival. The entire DS/SS modulation is usually below the carrier read-out noise level, but all information can be recovered with the local copy of the modulation code ("despreading") and error correction mechanisms [113]. Expected data transfer rates for direct laser communication between spacecraft are on the order of several tens of kbit/s. Bit error rates $< 26 \times 10^{-3}$ can easily be eliminated, e.g., with Reed-Solomon encoding [114].

4.3 NOISE SUPPRESSION SCHEMES

To remove excess noise and reach the required performance, one must apply additional correction schemes. This is necessary due to limitations in current analog-to-digital converter and reference oscillator technologies.

4.3.1 ADC JITTER CORRECTION

One of the very first components in the processing chain of the Phase Measurement System is the analog-to-digital converter which is triggered by a sampling clock. Intrinsic timing jitter within the ADC leads to a digitization of the input signal at non-equidistant intervals so that the digital replica is distorted. Figure 4.3 exaggerates this effect. One cannot distinguish between apparent phase shifts due to ADC timing jitter and a genuine gravitational wave signal. A timing stability requirement for the ADC on the order of a few tens of femtoseconds arises that depends on the maximum heterodyne frequency f_{het} (see Equation 35, Figure 3.22, and Table 9). Expressed in phase

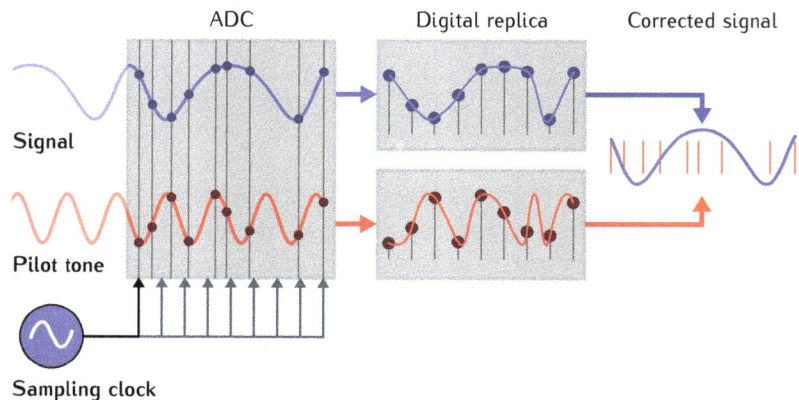

Figure 4.3: Even for a perfect system clock, the signal is digitized at non-equidistant intervals due to intrinsic timing jitter of the ADC. The digital replica of the signal will deviate from the original. To measure the intrinsic noise and remove it from the measured signal, a reference signal at different frequency (pilot tone) is digitized simultaneously by the same ADC.

noise, one would have to consider the actual sampling clock frequency f_s. The original requirement then scales by the ratio of f_s over f_{het} as

$$\left\langle \widetilde{\phi}_{sampling}^{\prime 10} \right\rangle (f) = \left\langle \widetilde{\phi}_{req}^{\prime 10} \right\rangle (f) \times \frac{f_s}{f_{het}} \stackrel{\wedge}{=} \left\langle \widetilde{t}_{req}^{\prime 10} \right\rangle . \tag{71}$$

This requirement is relaxed since the sampling frequency needs to be at least twice the maximum signal frequency (Nyquist–Shannon sampling theorem). However, the timing stability requirement remains the same no matter the actual signal frequency. Hence, for easier comparability when dealing with different signal frequencies, I will use data given in timing jitter instead of phase noise from now on.

All ADCs under test fail this requirement and would spoil the system performance if used as is. As illustrated in Figure 4.3, one solution to this problem is the introduction of a system-wide clean and stable reference signal (called "pilot tone"). This additional signal will be superimposed onto each ADC channel and hence is affected by the same ADC sampling time jitter. It is tracked by a dedicated DPLL core.

Since the pilot tone is well known we can now measure the ADC timing jitter. This information can then be used to correct the signal. Phase noise introduced by an unstable sampling clock is indistinguishable from ADC timing jitter and will be removed in the corrected signal by the same principles. Thus the pilot tone basically removes all timing stability requirements on the ADCs and the sampling clock.

There are several options to correct the signal (removing ADC timing jitter) all of which can be found in [115].

> We have to compare all measurements to this reference signal which needs to be the same constellation-wide. **Thus the timing stability requirements are now carried over to the oscillators producing the pilot tones** on the different spacecraft.

Unfortunately, as shown in Figure 4.4, oscillators that stable do not exist. In collaboration with Daniel Edler [116] we measured the phase noise

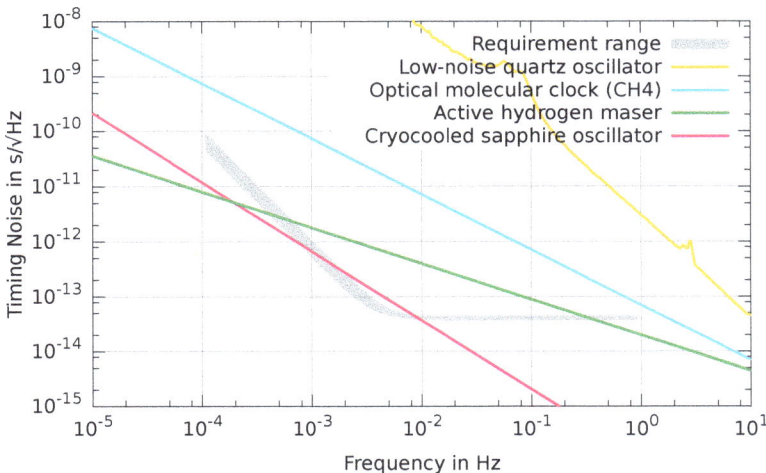

Figure 4.4: Rough trends of the timing stability over Fourier frequency for different reference oscillators. Even cryocooled dielectric-sapphire-resonator oscillators at 6 K (red) violate the requirement. A more realistic choice for a LISA reference oscillator would be closer to the yellow trace for a real measurement of a low-noise quartz oscillator.

Phase noise for electronic components is usually stated as dBc/Hz. At a given frequency offset from the carrier, this is the ratio of the noise power of a single sideband over the carrier signal, expressed in decibels in a 1 Hz bandwidth. This can be written in SSB phase noise (linear power spectral density) [117] and—considering the frequency it was originally referred to—converted to timing jitter for easy comparison.

of low-noise voltage controlled quartz oscillators ("VCXO_79.999#10" and "VCXO_80.001#A1" by *TSS microwave*) at 80 MHz. The performance of these devices is reasonably close to what to expect for an equivalent space-qualified low-power oscillator probably used in gravitational wave observatories. The noise exceeds the required level by orders of magnitude though.

Data for a HeNe/CH4-based optical molecular clock [118] and a commercial 100 kg Active Hydrogen Maser system ("10351" by *TIMETECH*) was extrapolated to the frequency range of interest – without finding a viable candidate. Even laboratory setups with state of the art cryocooled sapphire oscillators violate the required stability [119]. Noise information for all oscillators mentioned above was converted to timing jitter for comparison.

4.3.2 PILOT TONE JITTER CORRECTION

We have to deal with the excess phase noise of free-running oscillators that generate the pilot tones. The basic principle of the inter-spacecraft pilot tone jitter correction was already described in Section 2.2.2. The idea is to measure the differential phase noise between the pilot tones of the three different spacecraft and send this information back to Earth. Here it can be subtracted in post processing. Such an implementation requires the transmission of the pilot tone's phase noise information between the spacecraft. Figure 4.5 shows how this can be done by using the pilot tone to phase modulate the carrier wave of the outgoing laser beam with an electro-optic modulator (EOM). The resulting sidebands that hold the phase noise information of the respective pilot tones are sketched in Figure 4.6. After interference, the heterodyne signals now contain two additional beat notes (sideband beat notes) which—in parallel to the gravitational wave signal—yield the differential phase noise between the pilot tones.

The phase measured between local and remote sidebands is transmitted to Earth. Here, the raw information is processed by a hybrid-extended Kalman filter algorithm. The differential pilot tone jitter contained in the sideband beat notes will be used to synchronize all measurements performed on the

Figure 4.5: Remote and local pilot tones are phase modulated onto the laser beams by electro-optic modulators (EOM). The phase noise needs to be converted to achieve the required readout sensitivity.

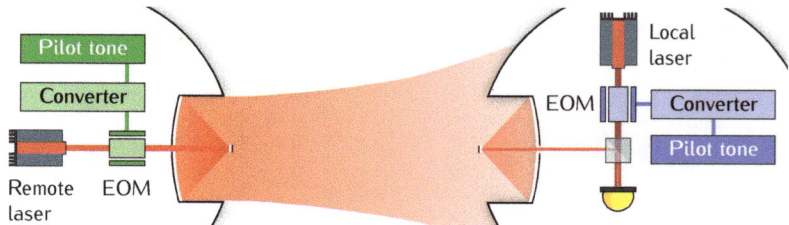

Figure 4.6: Sideband picture of the local (blue) and remote (green) laser beams. The upper and lower sidebands hold the phase noise information of the pilot tones.

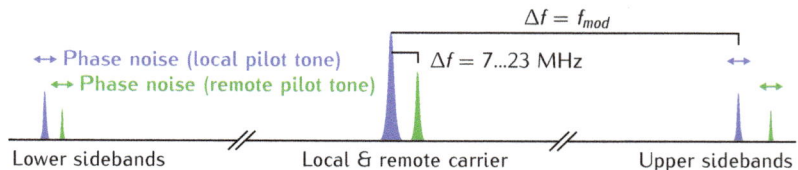

different spacecraft [91]. Additionally, information about the inter-spacecraft distances is recovered from the tracking information of the ranging codes. Subsequently, a Michelson interferometer of equal arm length and with a single virtual reference is constructed via time delay interferometry [88].

The sideband beat notes are separated from the carrier beat note by the difference in modulation frequency f_{mod}. Hence the frequencies of the phase modulation on the local and the remote spacecraft should differ by at least $\approx 1\,\text{MHz}$ so that DPLL cores can track the three beat notes individually. The 1 MHz shift in the reference oscillator frequency may shift the pilot tones between Phase Measurement Systems by $\approx 30\,\text{kHz}$, depending on the actual parameters used for metrology system. Yet this has no impact on the PMS design or any stability requirements. Doppler shifts between spacecraft affect the upper and lower sideband differently as the effect is proportional to the absolute light frequency. As a result, the sidebands are not entirely symmetrical but may be shifted up and down by up to $100\,\text{Hz}$. This has to be considered when designing the detailed beat note readout algorithms. For example, both sideband beat notes have to be treated separately due to the non-negligible difference in frequency.

Frequency dividers or multipliers convert between the pilot tone frequency and the actual modulation frequency. In contrast to the mixing process in, e.g., heterodyne interferometry (see Section 2.1.2) or electronic mixers, which maintain phase information, these devices do conserve timing jitter. This leads to a multiplication or division of phase noise by the respective frequency ratio as illustrated in Figure 4.7.

Original signal:

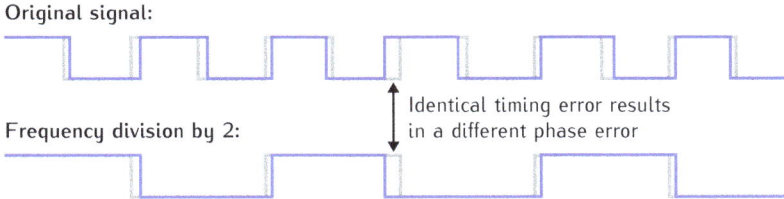

Frequency division by 2:

Identical timing error results in a different phase error

Figure 4.7: Noisy square wave signal (top, blue) with ideal representation (gray) and corresponding signals after frequency division. Phase noise is divided, timing jitter is conserved.

This conversion of phase shifts makes these devices ideal for scaling the pilot tones' phase noise. It potentially reduces the impact of sideband read-out noise in the equivalent displacement noise as explained in Section 2.2.2.1. The combined read-out of both sideband beat notes per heterodyne signal then results in a total equivalent displacement noise of

The read-out noise for both sidebands is uncorrelated, thus the **combined read-out** yields an improvement by a factor of $\sqrt{2}$. Equation 34 and requirements from Table 9 were used in the conversion.

$$
\begin{aligned}
\left\langle \tilde{x}_{\text{r/o}}^{\text{total}} \right\rangle_{\text{sidebands}} &= \frac{f_{\text{het}}}{f_{\text{mod}}} \frac{1}{\sqrt{2}} \frac{1}{J_1(m)^2} \left\langle \tilde{x}_{\text{r/o}}^{\text{total}} \right\rangle \\
&= \frac{f_{\text{het}}}{f_{\text{mod}}} \frac{1}{\sqrt{2}} \frac{J_0(m)^2}{J_1(m)^2} \left\langle \tilde{x}_{\text{r/o}}^{\text{total}} \right\rangle_{\text{carrier}} \\
&\overset{!}{<} \frac{f_{\text{het}}}{f_{\text{mod}}} \underbrace{\frac{1}{\sqrt{2}} \frac{J_0(m)^2}{J_1(m)^2}}_{= 9.37(\text{for } m = 0.53)} \times 10 \left\langle \tilde{x}_{\text{req}}^{/10} \right\rangle
\end{aligned}
\tag{72}
$$

which is only below the required level for modulation frequencies f_{mod} of at least 93.7 times the maximum heterodyne frequency f_{het}. To retain a strong carrier signal for the main science measurement only 7.5% of the respective carrier power is invested in each single sideband. This leads to a modulation index of $m = 0.53$ that determines the minimum modulation frequency (see Section 2.2.1.4). For this value, the modulation frequency needs to be in the lower GHz range. A higher modulation index would lower the required modulation frequency, but at the same time increase the overall influence of shot noise.

4.4 FREQUENCY DISTRIBUTION SYSTEM

The Inter-Spacecraft Frequency Distribution System is an essential part of the effort to synchronize all metrology systems between the different spacecraft. It generates all high fidelity reference signals for noise suppression. I concentrated my efforts on the design, development, and testing of this system, which also includes the evaluation of all components in the signal path locally and between the different spacecraft.

After a long run of pre-experiments described in Section 5.2 we decided to use a reference oscillator at a frequency of $f_{mod} = 2.40\,\text{GHz}$ for the phase modulation. This is above the minimum required modulation frequency for most considered mission concepts. Modulation frequencies between different laser links will differ by 1 MHz to get sideband beat notes at 1 MHz above and below the carrier beat note. The assumed heterodyne frequency range was $7\ldots23$ MHz which might only be possible with moderate improvements in laser relative intensity noise or photoreceiver electronics. A division of the modulation frequency by 32 generates a pilot tone with $f_p = 75\,\text{MHz}$. A $f_s = 80\,\text{MHz}$ sampling clock can be derived from a frequency division by 30 out of the same reference frequency. Thus the pilot tone is under-sampled and will be aliased down to 5 MHz. Including the sideband beat notes and the pilot tone, we assume a signal bandwidth of $5\ldots24$ MHz as illustrated in Figure 4.8.

This exact scheme was implemented on the current version of the developed Frequency Distribution System (see Section 5.3), but we can adapt it easily if necessary. In principle, beat notes are allowed to be as high as

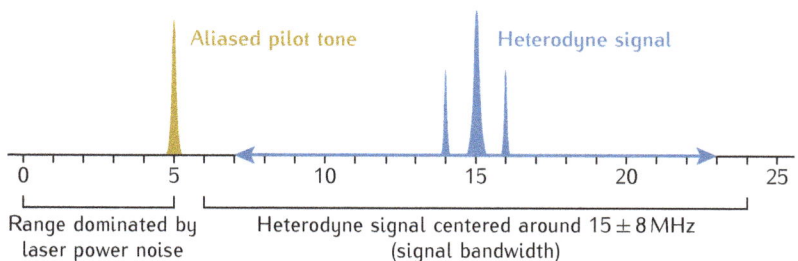

Figure 4.8: Down-aliased pilot tone at 5 MHz and heterodyne signals with sideband beat notes 1 MHz above and below the carrier which is centered around 15 MHz and shifted by 8 MHz due to Doppler shifts. Resulting signal bandwidth: $5\ldots24$ MHz.

$f_s/2 = 40$ MHz. If the frequency bounds do not include the lower frequency range due to relative intensity noise restrictions, a pilot tone at 37.5 MHz can be generated by a frequency devision of 64. This does not require any change in the sampling frequency but limits the maximum beat note frequency to 36.5 MHz [120].

If the pilot tone is significantly larger in amplitude than the heterodyne beat notes, it is possible to place it within the frequency range dominated by relative intensity noise after all. This would allow us to detect the pilot tone at 5 MHz. It might be beneficial though to amplify the beat notes in a way that they alone use up the entire dynamic range of the ADCs. In this case, the pilot tone would be limited to the maximum amplitude of the carrier beat note.

The current implementation allows for two **divide ratios**: 32 and 64. The divide ratio for the sampling clock can be set to any multiply of 2. A 3.2 GHz reference oscillator for instance divided by 40 would generate the same 80 MHz sampling clock. A possible pilot tone at 100 MHz or 50 MHz would then by aliased to 20 MHz or 30 MHz, respectively.

The final concept for the Inter-Spacecraft Frequency Distribution System is shown in Figure 4.9. The differential phase noise between the two 2.4 GHz reference oscillators can be measured via the sideband beat notes in a local reference interferometer and by electronically mixing both oscillator signals down to their 1 MHz difference frequency and subsequently passing them to the PMS. The phase noise of this signal will be read out via a dedicated DPLL core. With no optical signal present, the frequency range in this channel will not be dominated by relative intensity noise, so that high precision measurements become possible even for such low frequencies.

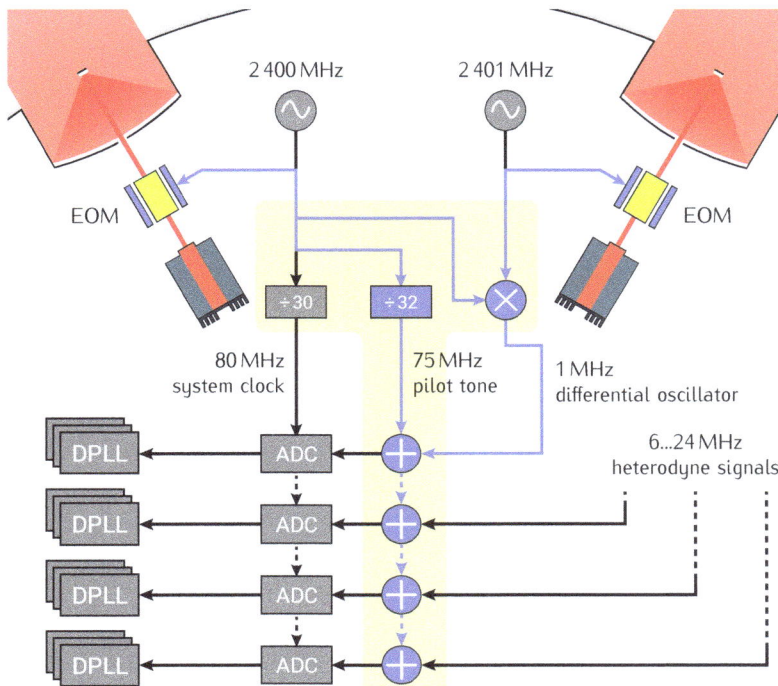

Figure 4.9: ADC and pilot tone jitter correction scheme. EOMs modulate the outgoing laser beams by reference oscillators at 2.4 GHz. This transfers their phase noise in sidebands to the distant spacecraft where it can be compared to the local oscillator's noise. Dividers downconvert the phase noise of the oscillator and produce the sampling clock and a pilot tone. The latter is added to each channel of the Phase Measurement System and will be used to correct for intrinsic ADC timing jitter. Multiple DPLL cores per channel track the different signal frequencies individually. Components and signal lines that must meet the timing stability requirement are highlighted in blue.

Figure 4.10: Simplified functional overview of the metrology system for one laser link. Quadrant photodiodes are used to enable Differential Wavefront Sensing (DWS) to not only measure position changes but also relative alignment information for the outgoing beam (detector A) and the proof-mass (detector B). This requires a multitude of phase measurement channels. Additionally, the metrology system provides alignment information with respect to the incoming beam for both, proof-mass and spacecraft, to the Drag-Free Attitude Control System (DFACS). All interactions between systems are shown as black arrows.

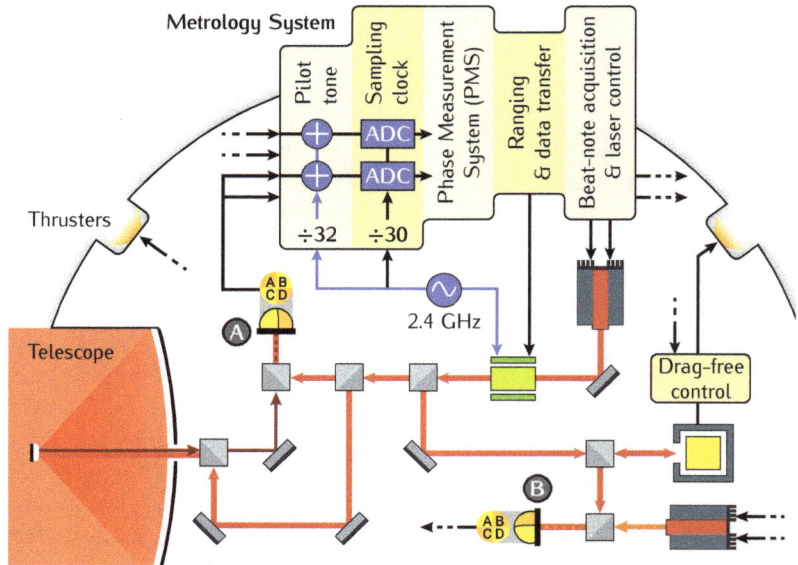

A functional overview of the metrology system for gravitational wave observatories for one laser link is shown in Figure 4.10. All optical components, e.g., mirrors and beam splitters, will be attached to an optical bench. There will be one of these benches per laser link. Light from the lasers and between the benches is transmitted via optical fibers.

> The timing stability requirement corresponding to the particular mission concept—that could not be met with ADCs and reference oscillators—**is now instead imposed upon all components of the pilot tone generation and transmission chain** highlighted in blue in Figure 4.9.

On the one hand this concerns all electrical cables [121], electro-optic modulators [76], optical amplifiers [122, 123], and any optical fibers [124] present in the GHz signal lines. This part is only necessary for the constellation-wide distribution between the spacecraft. On the other hand all components between the reference oscillators and the actual pilot tone—i.e., cables, dividers, adders, and filters—as well as the mixer that produces the difference frequency between both oscillators are subject to the same timing stability requirement. These components, which are highlighted in yellow in Figure 4.9, are part of the metrology system itself. Most of them are necessary even when considering local measurements only. ∎

FREQUENCY CONVERSION AND TRANSMISSION CHAIN

For a long time, the part of the Frequency Distribution System (FDS) that actually generates the different frequencies was treated as a black box as exemplified in Figure 5.1. We simply assumed it would output a sinusoidal GHz signal stable in phase with respect to a reference frequency. This was sufficient though to evaluate the different electrical, electro-optical, and optical components in the gigahertz signal path from the FDS to the outgoing laser beam (highlighted in orange, see Section 5.1). We later dealt with the frequency conversion itself (Section 5.2) before turning our attention to the details of the megahertz signal path (highlighted in green, see Section 5.3).

When I speak of the Frequency Distribution System, I refer to the frequency converter electronics (black box) including the components of the MHz signal path (green) which are all a part of the overall metrology system. The full pilot tone generation and transmission chain additionally includes the GHz signal path—for inter-spacecraft distribution—with all electro-optical and optical components (orange). None of the components in either signal path is allowed to add excess timing jitter above the required level $\left\langle \tilde{t}_{\mathrm{req}}^{/10} \right\rangle$ as stated in Table 9 on page 86. The same is true for the frequency conversion process since the differential timing jitter between the MHz and the GHz signal must fall below the same noise level. Only then all measurements throughout the constellation can be synchronized sufficiently well.

Figure 5.1: Frequency converter as 'black box' and simplified GHz and MHz signal paths (orange, green) with different components like cables, electro-optic modulators (EOMs), and fiber amplifiers (FAs).

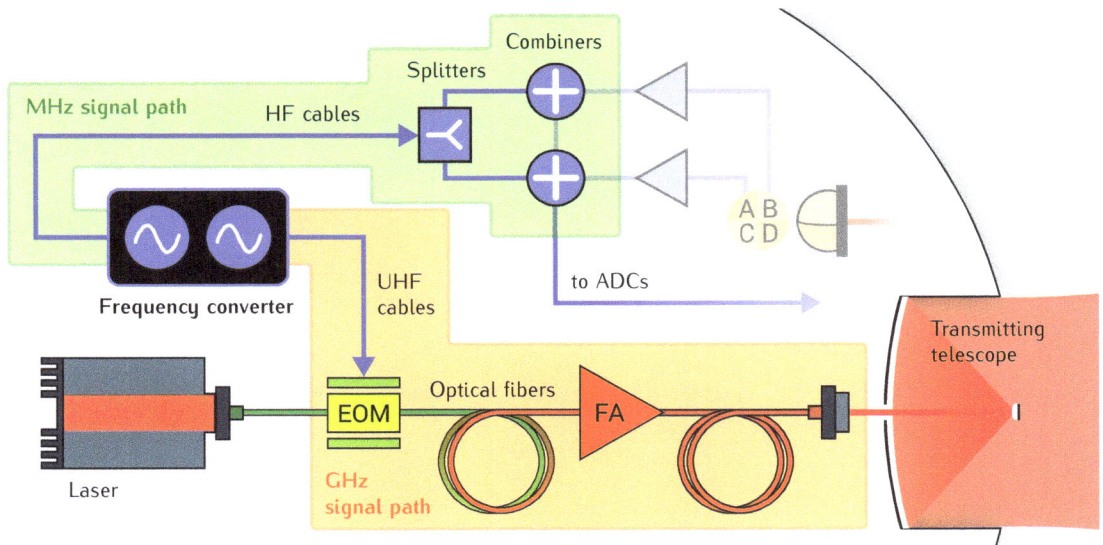

5.1 GIGAHERTZ SIGNAL PATH

The gigahertz signal is transmitted to the electro optic modulator (EOM) via ultra high frequency (UHF) cables and carried along the long arms together with the laser carrier as GHz sidebands. It passes optical fibers and laser fiber amplifier (FA) stages before being sent to the distant spacecraft.

5.1.1 ULTRA HIGH FREQUENCY CABLES

Electrical UHF cables transport the GHz signal from the frequency converter to the EOM. The transmission of ultra high frequency signals is a science of its own. After initial research on a number of cables it became clear that standard UHF cables that use a dielectric layer made of polytetrafluoroethylene (PTFE) or "Teflon™" by *DuPont* may not be suitable for gravitational wave observatories. This is due to the fact that PTFE changes its crystalline structure from the helix phase II to the hexagonal phase IV at about room temperature [125]. This leads to a change in dimension and density. The influence of this effect on the phase stability of cables is well known [121, 126]. The density change in the dielectric layer leads to a change in the velocity of propagation and thus results in a rapid change of the electrical length of the cable. This effect is referred to as the 'Teflon knee'. The relationship between electrical length and temperature is largest at room temperature ($15\ldots25°C$) and can increase, compensate, or even reverse the physical length change of the cable. Since we have to consider that the spacecraft will be designed to have an inside temperature similar to this range, the real extent of this effect is of general concern. In the following I will present detailed measurements of the magnitude of the phase stability over temperature that cover the wider area around the 'Teflon knee'.

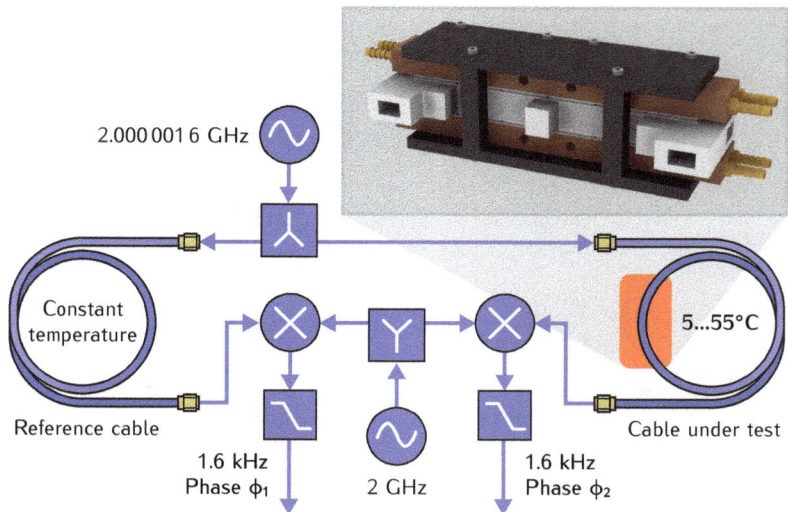

Figure 5.2: Test stand to measure phase temperature coefficients of UHF cables: A 2 GHz signal is split and passed through the cable under test and a reference cable of equal design and length. The phase of both signals is measured after mixing them down to a more convenient frequency of 1.6 kHz. The device that heats and cools a 28 cm section of the cable under test in a range of $5\ldots50°C$ is shown in the upper right corner.

In collaboration with Amrit Pal Singh [127] we developed a test stand (Figure 5.2) that changes the temperature of a cable slowly (period of hours) repeatedly over the course of some days. A 2 GHz signal was split and passed through the cable under test and a reference cable of equal design and length, respectively. The phase of both signals was measured with a precision of < 0.1 mrad after mixing them down to a more convenient frequency. We thermally connected a 28 cm section of the cable under test by metal foam to a unit of 12 water-cooled Peltier elements. Up to 600 W of heating or cooling power reached an effective temperature range of 5 . . . 50°C. This design, illustrated in Figure 5.2 top right corner, is the result of many iterations. It provides a wider temperature range than previous designs and is compatible even with very short cables.

To analyze the magnitude of the differential phase change $\Delta\phi$ of an electronic signal passing through the cable under test due to a temperature change ΔT at a given temperature T, we developed a specialized computer algorithm. It automatically computes phase stability values in units of radians per Kelvin and meter. The computation of the coefficients assumes a linear relationship between phase stability and signal frequency f. Although research exist indicating nonlinear effects at frequencies between 5 and 100 MHz [128], our measurements show that this assumptions seems to be valid with frequencies between 500 MHz and 2 GHz—at least for the cables we tested. Thus we are allowed to convert the coefficients to an equivalent change in signal arrival time per meter cable over temperature, independent of the signal frequency:

> Earlier experiments at the *Albert Einstein Institute* were limited to a temperature range of 25 . . . 45°C that prohibited the observation of the 'Teflon knee' [76].

$$\underbrace{\frac{\Delta\phi}{\Delta T}\left[\frac{\text{rad}}{\text{K}}\right]}_{\substack{\text{per meter cable,}\\\text{valid at 2 GHz}}} / \underbrace{(2\pi \times 2\,\text{GHz})}_{\text{see Equation 35}} = \underbrace{\frac{\Delta t}{\Delta T}\left[\frac{\text{s}}{\text{K}}\right]}_{\text{per meter cable}} . \tag{73}$$

This was done for Figure 5.3. As an example, five relevant cables of three different types are shown. The width of each trace shows the range of the calculated coefficients that is different for cooling and heating periods. As

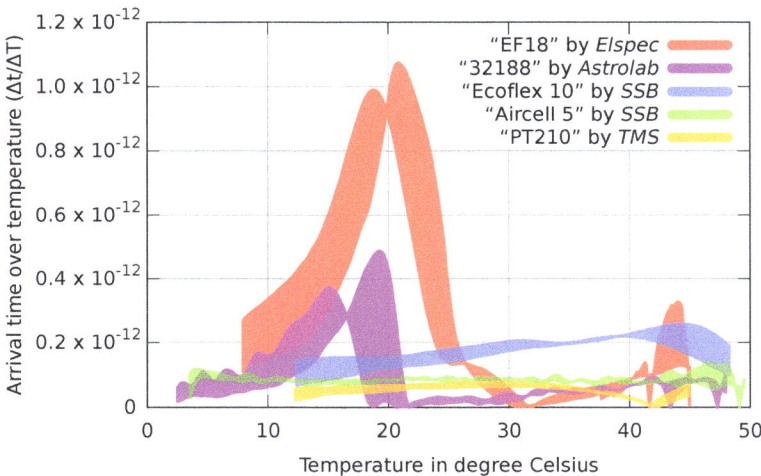

Figure 5.3: Timing stability coefficients $\Delta t/\Delta T$ over absolute temperature for 5 different cables: The width of each trace shows the range of the coefficients that is different for cooling and heating periods.

expected, the two PTFE-based cables ("EF18" by *Elspec* and "32188" by *Astrolab*, now *HUBER+SUHNER*) show a distinct maximum between 15 and 25°C. Both cables use a low-density PTFE dielectric layer to improve the phase stability, yet the peak values in the temperature range of interest are at 1.2×10^{-12} s/K (red) and 4.9×10^{-13} s/K (purple), respectively. These values translate to temperature stability requirements of

$$\underbrace{\left\langle \widetilde{T}_{\text{req}}^{\prime/10} \right\rangle \left[\frac{\text{K}}{\sqrt{\text{Hz}}} \right]}_{\text{per meter cable}} = \left\langle \widetilde{t}_{\text{req}}^{\prime 10} \right\rangle \left[\frac{\text{s}}{\sqrt{\text{Hz}}} \right] \times \frac{1}{\Delta t/\Delta T} \left[\frac{\text{K}}{\text{s}} \right] . \tag{74}$$

Typical values fall between 10^{-2} and $1\,\text{K}/\sqrt{\text{Hz}}$ per meter and are summarized for all five cables under test in Figure 5.4.

To avoid the Teflon knee, the "PhaseTrack210" cable by *Times Microwave Systems* uses a proprietary dielectric material under brand name TF4™. Although quite expensive, the stability of these cables is extremely good over the entire temperature range with a coefficient of less than 8.0×10^{-14} s/K (yellow). Another type of cables uses a low-density polyethylene (PE) dielectric that shows a phase stability more constant over temperature. We present measurements for two different cables, "Ecoflex 10" and "Aircell 5" by *SSB-Electronic*. While the minimum coefficients never reach the level of PTFE-based cables at high temperatures, the maximum coefficients are well below the level of PTFE cables with 2.6×10^{-13} s/K (blue) and 1.4×10^{-13} s/K (green), respectively. PE-based cables are generally not equipped for frequencies above 10 GHz, but are considerably cheaper than all cable types mentioned above.

> These **low-density PE** cables should not be confused with standard PE cables which use a a solid dielectric layer and feature coefficients of 10^{-12} s/K and larger.

UHF cables differ in more than just the dielectric material and phase stability though. Other important specifications are, e.g., the operating temperature range, the maximum supported frequency, and the velocity of propagation (VoP) in fractions of the speed of light. Table 10 compares the different features.

> **Table 10:** Specifications of the cables under test including the measured maximum timing stability coefficients per meter cable.
>
> *Cables from *Astrolab* and *Times Microwave Systems* are also available as space qualified equivalents.

Cable name (manufacturer)	Dielectric material	Operating temperature	Max. frequency	VoP	Coefficient $\Delta t/\Delta T$ [s/K]
EF18 (Elspec)	low-density PTFE	−40…85°C	18 GHz	77%	1.2×10^{-12}
32188 (Astrolab)*	low-density PTFE	−55…200°C	27 GHz	86%	4.9×10^{-13}
Ecoflex 10 (SBB)	low-density PE	−55…85°C	6 GHz	85%	2.6×10^{-13}
Aircell 5 (SBB)	low-density PE	−55…85°C	10 GHz	82%	1.4×10^{-13}
PhaseTrack210 (TMS)*	proprietary TF4™	−55…150°C	29 GHz	83%	8.0×10^{-14}

Given the right absolute temperature, the coefficients that describe the arrival time over temperature change can be considerably lower. Additionally, in laboratory setups cables often introduce common mode noise that is automatically suppressed in differential measurements and only unmatched cable lengths are of concern. For gravitational wave observatories however,

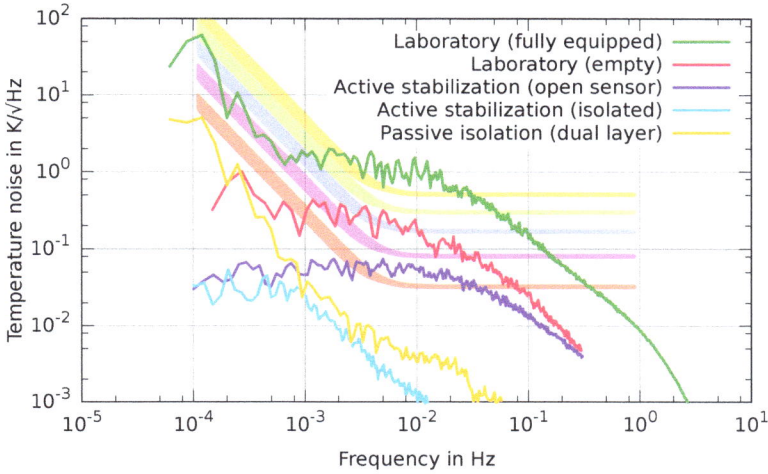

Figure 5.4: Temperature stabilities vs. timing stability coefficients of UHF cables converted to frequency dependent timing stability requirements valid for a cable length of 1 m. Colors relate to the cables shown in Figure 5.3. Temperature stabilities measured in a laboratory at the *Albert Einstein Institute* in Hanover, Germany. Exemplary values achievable with passive isolation and active temperature stabilization are shown.

common mode noise suppression should be a subject of a more detailed study that involves a full time delay interferometry simulation. Most probably, the entire signal path from the frequency converter to the EOM has to be considered.

The situation is not as dire as implied though. Temperature stability on board the spacecraft is a high priority in any case due to requirements of the optical bench (see Section 2.1.3). Even in laboratories on ground, environments sufficiently stable in temperature can be created, as shown in Figure 5.4. Here, all 'worst-case' coefficients converted to frequency dependent timing stability requirements valid for a cable length of 1 m are summarized and compared to actual temperature stabilities. The data was taken in laboratories at the *Albert Einstein Institute* in Hanover, Germany. Active equipment, passive isolation, and active temperature stabilization for the measurement setup of concern can change the stability dramatically. The main message of the plot though is that UHF cables are a potentially limiting noise source. If the wrong type of cable is chosen or temperature fluctuations are not kept under control, even a one meter cable can spoil the entire performance of a measurement setup.

5.1.2   ELECTRO-OPTIC MODULATORS

Electro-optic modulators (EOMs) house a crystal such as lithium niobate ($LiNbO_3$) that changes its refractive index – and thereby the optical path length – linearly in proportion to the strength of a local electric field. This shifts the phase of light passing through the crystal. The effect is known as electro-optic effect (or Pockels effect). Oscillating phase shifts produce sidebands on the laser carrier that are stable in phase with respect to the modulation signal. Hence we can apply the GHz signal directly to the EOM to generate sidebands that carry the pilot tone's phase information to the distant spacecraft. First-order sidebands appear at the modulation frequency, higher order sidebands at multiples of it. The amplitude of the oscillating phase shift is usually described as modulation index m in units of radians. The power of carrier and sidebands can be calculated by Bessel functions of the first kind for the value m [129]. For a targeted power of 7.5% in a single sideband with respect to the carrier, a modulation index of $m = 0.53$ is needed (see Section 2.2.1.4).

There are a number of effects that can degrade the phase fidelity between modulation signal and sideband: a change of the electric potential, a change of the refractive index due to different optical powers, physical length changes, and a change of the refractive index due to temperature fluctuations. An environmental temperature change affects the EOM just as absorbed optical or electrical power does [76].

The EOM used in gravitational wave observatories has to take care of both, the GHz sidebands and the DS/SS modulation for ranging and data transfer. Thus broadband modulators are required. The only viable choice are fiber-coupled EOMs with an integrated optical waveguide as they feature about five times higher efficiency than free-beam EOMs. It is important to know if 1) fiber-coupled EOMs are phase stable enough under realistic conditions, and 2) what the required electrical power is to reach the desired modulation index of $m = 0.53$. We evaluated two different EOMs.

JENOPTIK   The commercially available polarization-maintaining "Integrated Optical Phase Modulator" by *Jenoptik* allows broadband phase modulation for frequencies up to 3 GHz. Its waveguide is made of magnesium oxide doped lithium niobate ($MgO:LiNbO_3$) which provides a high damage threshold and can handle up to 300 mW of optical power. The maximum voltage was specified as 40 Vpp. It comes assembled with polarization-maintaining single-mode fibers and FC/APC connectors. A typical overall insertion loss of 4 dB was observed.

LITEF   The custom built "PM1064" EOM by *Northrop Grumman LITEF GmbH* (Figure 5.5) is not available off-the-shelf. It was designed to comply with space applications, is radiation hard and vacuum compatible. The functional principle is identical to the *Jenoptik* EOM, although this one allows for modulation frequencies of up to 10 GHz but features a lower damage threshold of only 50 mW. It came with bare fibers and we spliced FC/APC connectors to the EOM manually.

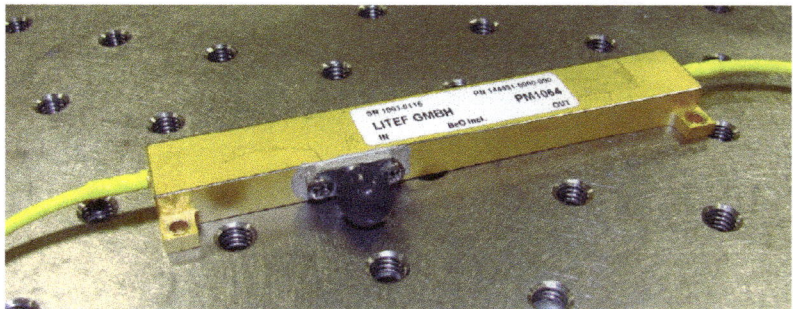

Figure 5.5: Radiation hard and vacuum compatible "PM1064" EOM by *Northrop Grumman LITEF GmbH*.

5.1.2.1 PHASE FIDELITY

To measure the excess phase noise that an EOM adds to a sideband in relation to the modulation signal, we used a setup presented as simplified sketch in Figure 5.6. The EOM modulated laser light with a frequency of $f_{mod} = 2\,\text{GHz}$ $+1.6\,\text{kHz}$. We evaluated interference signals after combining the modulated light with a second laser beam. Both lasers were phase locked with an offset frequency of $f_{ref} = 2\,\text{GHz}$. The difference between both frequencies amounted to 1.6 kHz, hence the detectable beat notes were at

✦ 1.6 kHz (beat note between reference carrier and upper sideband) referred to as sideband beat note,

✦ 2 GHz (beat note between both carriers) referred to as carrier beat note, and

✦ 4.000 001 6 GHz (beat note between the reference carrier and the lower sideband) which is not needed.

These beat notes can be derived from the carriers and sidebands present in the heterodyned laser beam as illustrated below the setup.

The carrier beat note was mixed down with the modulation frequency which gave us a second signal at 1.6 kHz. It can be shown that most noise sources in such a setup are common mode between both signals, yet only the sideband beat note holds the information about the EOM phase noise [76]. Thus the differential phase noise between both signals is an upper limit on the excess noise of a single sideband introduced by the EOM. It additionally

Figure 5.6: Setup for measuring the phase characteristics of a single EOM sideband at GHz frequencies: Two lasers were offset phase locked to a reference frequency with one beam sent through an EOM driven by a GHz modulation signal. Both beams were heterodyned, and sent to a fiber-coupled photodiode. The carrier-carrier beat note was mixed down and low-pass filtered. Subsequently the phase of the resulting signal was measured alongside the phase of the sideband-carrier beat note. The carriers and sidebands present in the heterodyned laser beam are illustrated below. Details can be found in [76].

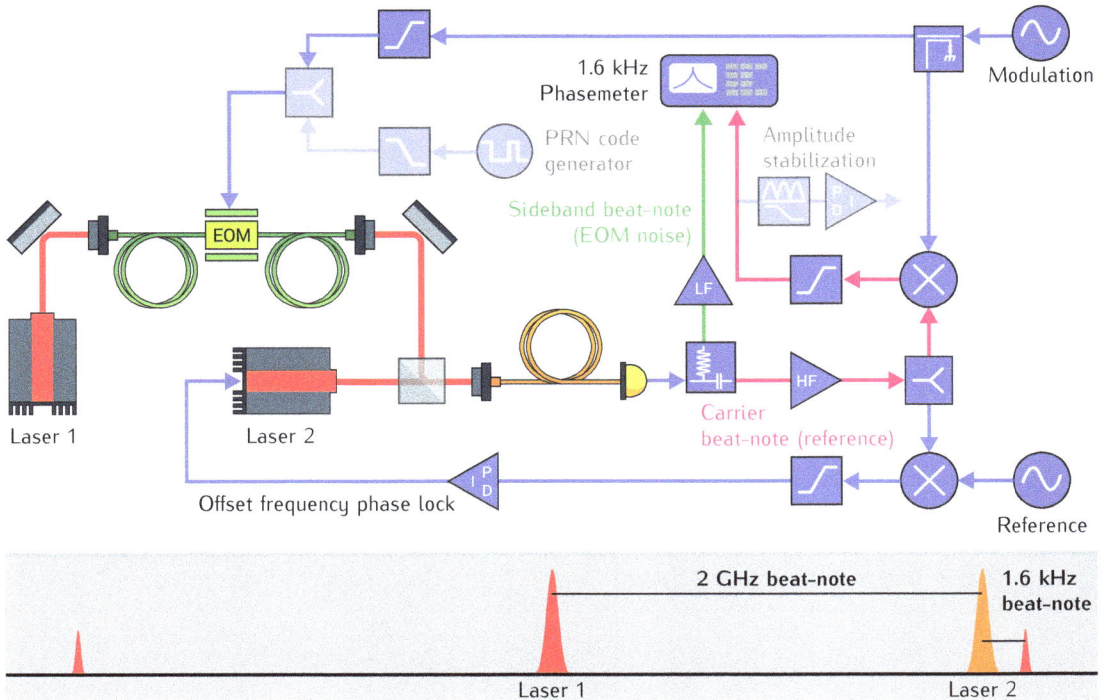

comprises phase noise from, e.g., the UHF cable carrying the modulation frequency.

As it turns out, one critical and potentially limiting noise source is the mixer (double-balanced Level 7 coaxial mixer ("ZX05-C24-S+" by *Mini Circuits*) in the carrier beat note path. It spuriously translates fluctuations in the signal's amplitude to a phase shift. The magnitude of this effect changes non-linearly over absolute signal power (local oscillator and RF input). To overcome this noise source, we stabilized the power of the carrier beat note. Amplitude stabilization electronics were used as described in [124] that acted on the laser diode current. Alternatively, when possible, it can be sufficient to fine-tune the laser power incident on the photodiode so that the coupling factor from amplitude to phase change was at a minimum.

We could show that both, the *Jenoptik* and the *LITEF* EOMs, have a sufficient phase fidelity and reach the stringent timing stability requirements [65, 124]. In addition, we combined the 2 GHz +1.6 kHz modulation signal with a realistic PRN code (DS/SS modulation) to reveal the impact of the PRN code and the additional combiner electronics in the GHz signal path. These electronics are: 1) a standard off-the-shelf DC to 10 GHz power combiner ("ZX10R-14+" by *Mini Circuits*), and 2) filters (1.9 to 2.7 GHz high pass filter "VHP-16" and DC to 825 MHz low pass filter "VLFX-825" by *Mini Circuits*) to prevent the modulation signal to arrive at the PRN code generator and vice versa. With all these components and the additional DS/SS ranging modulation present, the overall phase noise was still found to meet the required level with a comfortable margin.

A separate study was performed to evaluate the impact of the combiner electronics and the modulation sidebands to the **ranging accuracy** under weak light (one laser attenuated to 100 pW) conditions. It was found that there is neither a decrease in accuracy nor an increase in bit error rate [130].

All results are summarized in Figure 5.7, converted to timing jitter. Measurements were performed at a 2 GHz modulation. Although a linear increase in phase noise is expected with increasing frequencies, the conversion to timing jitter removes this frequency dependence. Thus the expected timing stability is the same and this noise level should be applicable for dif-

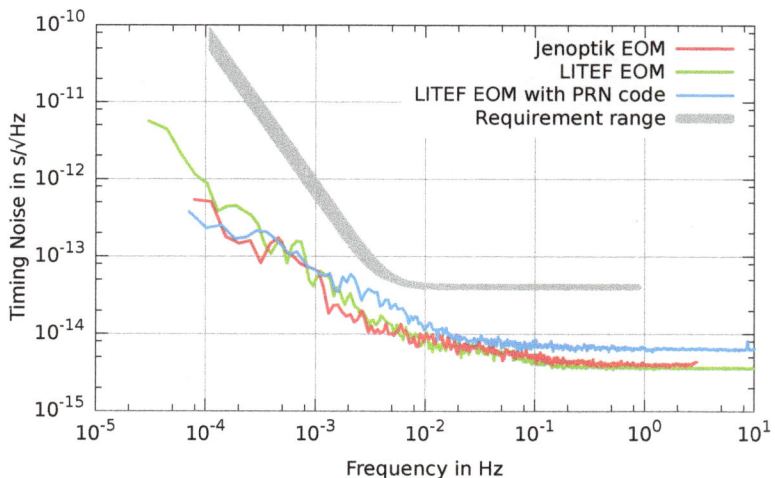

Figure 5.7: The *Jenoptik* and the *LITEF* EOM both have a sufficient phase fidelity and match the stringent timing stability requirements. Measurements were performed at 2 GHz modulation frequency.

ferent frequencies of the same order of magnitude. A significantly higher frequency is only required for much lower modulation indices which would at the same time tighten the timing requirements slightly. This is due to the lower carrier read-out phase noise for the stronger carrier power at smaller modulation depths. However, the marginal improvement in sensitivity might not be worth the effort.

At present, we assume a light power of 7.5% of the carrier power for each modulation sideband. This requires a modulation frequency that is 93.7 times the maximum heterodyne frequency (see Equation 72). Reducing the sideband power by one order of magnitude results in an increase in the required modulation frequency by roughly a factor of 10 (up to 26 GHz). Yet, it decreases the carrier read-out noise by less than 15%. If anything, a small increase in sideband power might be advisable if it helps to simplify certain electronic components in the pilot tone generation and transmission chain.

For modulation indices $m < 1$ (no higher order sidebands) the sideband power is inversely proportional to the frequency factor.

The carrier light power can at maximum increase by 15% if no modulation sidebands are present.

5.1.2.2 TEMPERATURE

Basically no coupling between laser power and EOM induced phase shift can be found [76]. Thus part—if not most—of the observed phase noise caused by the EOMs is due to environmental temperature fluctuations. In fact, many passive thermal isolation layers were necessary to show the presented phase fidelity in a laboratory setup.

We found that the *Jenoptik* EOM requires a laser power stability better than $125 \, \text{mW}/\sqrt{\text{Hz}}$. The power stability of gravitational wave observatories is assumed to be orders of magnitude better [124].

To reveal the coefficients describing the phase shift over temperature change, the EOMs were thermally connected to Peltier elements as illustrated in Figure 5.8. A maximum heating power of almost 70 W was available.

The *Jenoptik* EOM was exposed to temperatures between 22 and 25°C. We found a linear dependency with a coefficient of 1.6×10^{-13} s/K [124] that is similar to 1 meter of "Aircell 5" cable (see Section 5.1.1). By that, a realistic temperature stability of 5×10^{-2} K/$\sqrt{\text{Hz}}$ at 10^{-2} Hz (see Figure 5.4) translates to an expected timing jitter of 8×10^{-15} s/$\sqrt{\text{Hz}}$. This is very close to the observed noise level of the *Jenoptik* EOM at the given Fourier frequency.

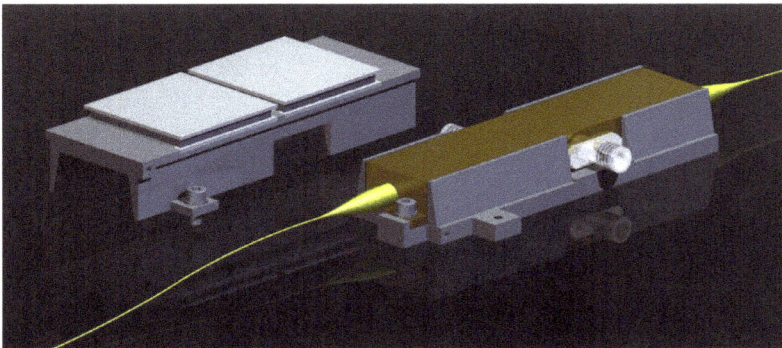

Figure 5.8: EOM thermally connected to a housing with two Peltier elements that provides a maximum heating power of almost 70 W.

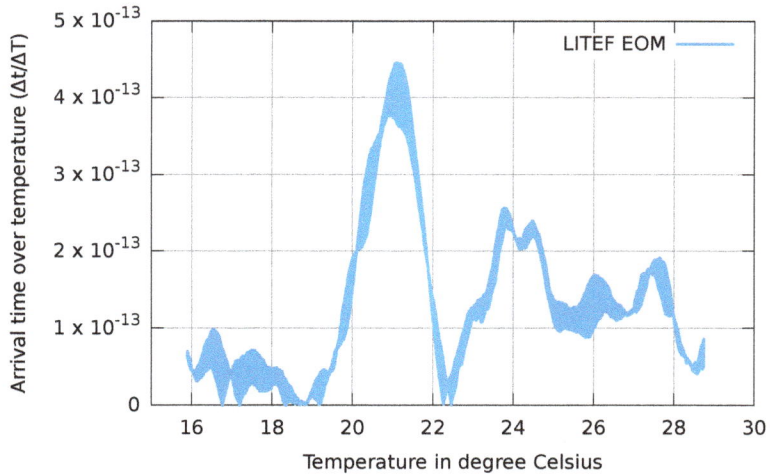

Figure 5.9: Timing stability coefficients $\Delta t / \Delta T$ over absolute temperature for the *LITEF* EOM: The width of each trace shows the range of the coefficients that is different for cooling and heating periods. Measurement performed at 2 GHz.

The *LITEF* EOM was examined for a wider temperature range between 15 and 30°C. The tools developed for the evaluation of phase stability coefficients of UHF cables was used for the measurement [127]. Figure 5.9 shows a correlation that is dependent on absolute temperature and peaks at 21°C with a value of 4.5×10^{-13} s/K. The correlation is lower by a factor of 2 to 5 in the remaining temperature range. The average value in the range of $22 \ldots 25$°C is almost identical to the correlation found for the *Jenoptik* EOM in the same limited temperature range.

This result does not only stress the importance of strict temperature stability requirements for the electronics and laser systems on board the spacecraft. It also implies that **the absolute temperature should be fine tuned to a level where the phase noise coefficients of the different components are at a minimum.**

5.1.2.3 EFFICIENCY

The required power to reach a certain modulation index m at a distinct frequency differs between both investigated EOMs as they were designed for a different frequency range. Assuming a change of the refractive index linearly proportional to the strength of the local electric field, the modulation index—which is the amplitude of the phase modulation—should scale linearly with the applied voltage. For a frequency range where the EOM is impedance matched to $50\,\Omega$, one can calculate a relationship between m and the electrical power in the modulation signal, P_{mod}, to be

$$m = \eta\left(f_{\mathrm{mod}}\right) \times \sqrt{P_{\mathrm{mod}} \times 50\,\Omega}\,, \tag{75}$$

with $\eta(f_{\mathrm{mod}})$ being the efficiency of the EOM for a modulation frequency f_{mod} [76].

The modulation index can be calculated when the power of the carrier, P_{carrier}, and of a single sideband, P_{sideband}, are known. The relationship between the three values is described by

$$\frac{P_{\mathrm{sideband}}}{P_{\mathrm{carrier}}} = \left(\frac{J_1(m)}{J_0(m)} \right)^2 . \tag{76}$$

Within the linear range, the output signal of the scanning Fabry-Perot's photodetector is proportional to the optical power of the carrier and sidebands as described in detail in [76]. Using this, we determined the sideband-over-carrier power for a sufficient number of modulation powers for both EOMs. Different modulation frequencies were used for the measurements. The *Jenoptik* EOM was evaluated at 400 MHz and 2 GHz while the *LITEF* EOM was set to a number of different frequencies, including 1 GHz, 2.5 GHz and 8 GHz. The results fit the predicted behavior quite nicely (see Figure 5.10) and we calculated efficiencies for both EOMs at the different frequency settings. The *Jenoptik* EOM features efficiencies of

$$\eta(0.4\,\mathrm{GHz}) = 0.67\,\frac{\mathrm{rad}}{\mathrm{V}} \quad \text{and} \quad \eta(2.0\,\mathrm{GHz}) = 0.37\,\frac{\mathrm{rad}}{\mathrm{V}} \tag{77}$$

while the *LITEF* EOM is more efficient even at higher frequencies with

$$\eta(2.5\,\mathrm{GHz}) = 0.65\,\frac{\mathrm{rad}}{\mathrm{V}} \quad \text{and} \quad \eta(8.0\,\mathrm{GHz}) = 0.5\,\frac{\mathrm{rad}}{\mathrm{V}} . \tag{78}$$

From the many efficiencies measured for the *LITEF* EOM at fine frequency intervals between 1 and 13 GHz, a plot of the efficiency over frequency could be derived. This data is shown in Figure 5.11. In conclusion, the modulation signal needs to have an electrical power of roughly 16 dBm to reach a modulation index of $m = 0.53$ (7.5% power with respect to the carrier in each

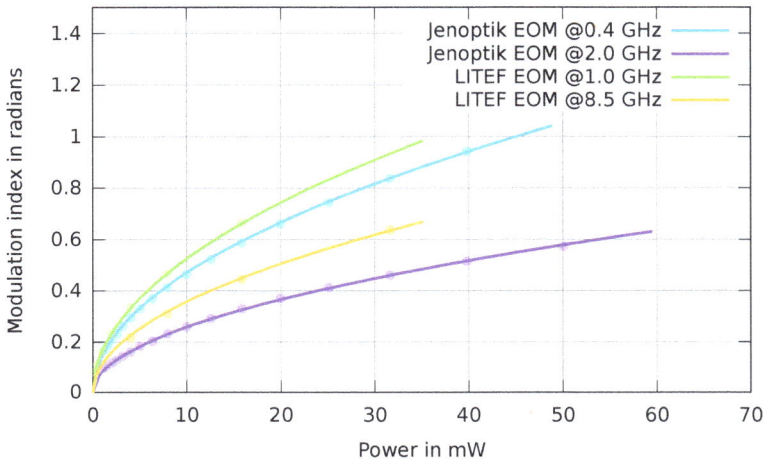

Figure 5.10: Modulation index over modulation signal power for the *Jenoptik* and the *LITEF* EOM at different modulation frequencies. The traces show a fit that reveals the efficiency of the EOM under test at this particular frequency.

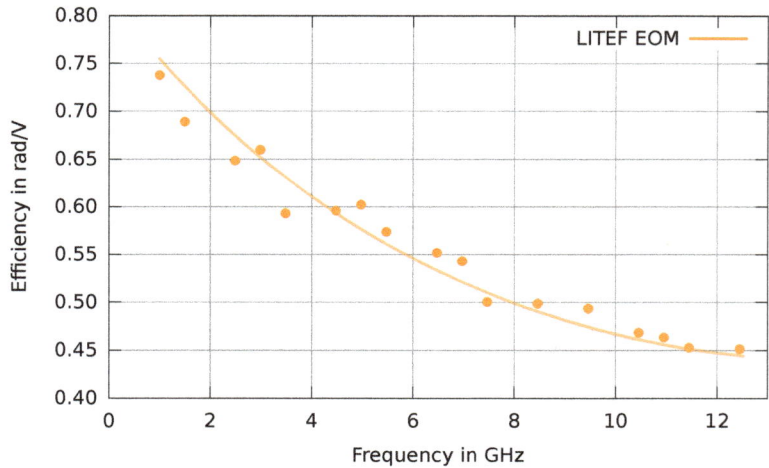

Figure 5.11: Efficiency over modulation frequency for the *LITEF* EOM: the efficiency of the *LITEF* EOM was determined for a total of 17 different modulation frequencies. The data reveals a not clearly described correlation.

sideband) at 2 GHz when dealing with the *Jenoptik* EOM. For the same modulation depth, the *LITEF* EOM requires less than 12 dBm electrical power. For 8 GHz frequency however, the required power rises to 21.5 dBm.

5.1.3 OPTICAL FIBERS

Optical fiber cables change their index of refraction and physical dimensions with temperature. This is generally uncritical since it only adds to the already dominating laser frequency noise that we have to deal with anyway (see Section 2.2.5). However, this effect is frequency dependent, and—with sidebands present in the laser signal—it causes a phase shift between sidebands and carrier.

We determined the magnitude of the phase shift over temperature change for two different optical fibers for a modulation frequency of 2 GHz. A technique similar to the UHF cable measurement was used to heat and cool a section of a fiber that replaced the EOM in the measurement setup described in Section 5.1.2.1. The determined coefficients are constant over a temperature range of $22 \ldots 29°C$. According to our measurements, a 1 m section of the fiber connected to the *Jenoptik* EOM's output has a thermal coefficient of 6.3×10^{-14} s/$\sqrt{\text{Hz}}$. The same section of a *Schäfter + Kirchhoff* single-mode polarization maintaining fiber cable is more stable with 4.6×10^{-14} s/$\sqrt{\text{Hz}}$ [124].

These coefficients are not in agreement with the value predicted theoretically for bare fibers [131]. Yet the discrepancy can be explained by the fibers' individual jacketing. The coefficient for a bare fiber was found to be about four times smaller than for a fiber with nylon jacketing. Additionally, the different jacketing of the two fibers can explain the observed difference in stability. However, both coefficients are much smaller than for any cable under test and should not limit the sensitivity of our observatory.

The jacketing of the fiber attached to the *Jenoptik* EOM was much stiffer and harder than the one of the *Schäfter + Kirchhoff* fiber.

5.1.4 FIBER AMPLIFIERS

The damage threshold of current fiber-coupled EOMs is the reason why a separate laser amplification stage is required that is situated behind the EOM. Thus it is in the GHz signal path and—similar to passive optical fibers—may add phase noise between the carrier and the modulation sidebands due to a number of effects like, e.g., optical path length fluctuations due to changes in the physical length or refractive index or non-linear dispersion.

The measurement setup used for EOM phase fidelity measurements can be adapted to evaluate fiber amplifiers as illustrated in Figure 5.12. Here we evaluated the interference signal of light from two lasers before and after the light passed through the fiber amplifier under test. Both lasers were phase-locked with an offset frequency of $f_{mod} = 2\,\mathrm{GHz}$ so that one laser simulated a single sideband at this modulation frequency. The beat note was mixed down with a reference frequency $f_{ref} = 2\,\mathrm{GHz} + 1.6\,\mathrm{kHz}$ and we measured the phase of the resulting 1.6 kHz signals. The differential phase holds the information on excess phase noise introduced by the fiber amplifier to the simulated sideband with respect to the carrier. The laser power incident on the two photodetectors was stabilized by actuating on the beam alignment into the fiber coupler with a piezoelectric mirror using the respective beat note amplitude as a reference.

Many different laser systems and optical amplifiers have been investigated over the years [66, 106, 107, 123]. While viable concepts were identified for output powers of up to 1 W, stimulated Brillouin scattering seems to increase the phase noise between carrier and sidebands above the required level for higher output powers [122]. As an example, phase noise levels for an

Figure 5.12: Setup for measuring the phase fidelity of a fiber amplifier at GHz frequencies: Two lasers were offset phase locked to an offset frequency so that one laser carrier simulates a single sideband. Both beams were heterodyned and part of the beam was sent to a fiber-coupled photodiode (reference beat note). The other part was passed through the fiber amplifier under test and subsequently sent to an identical photodiode (signal beat note). Both beat notes were mixed down and low-pass filtered. Subsequently we measured the phase of the resulting signals. The carriers present in the heterodyned laser beam are illustrated below. Details can be found in [122].

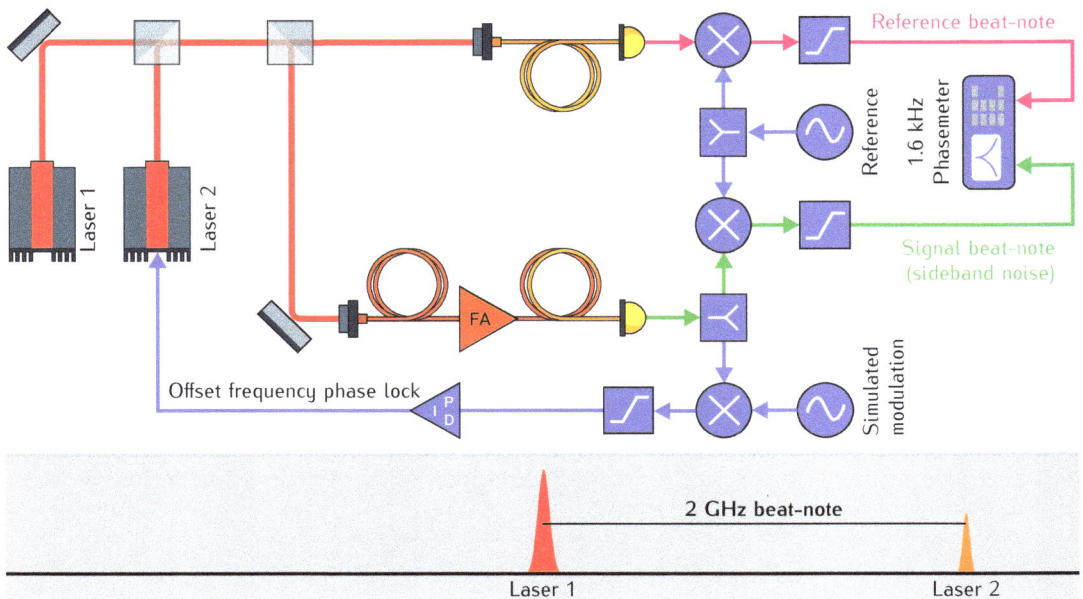

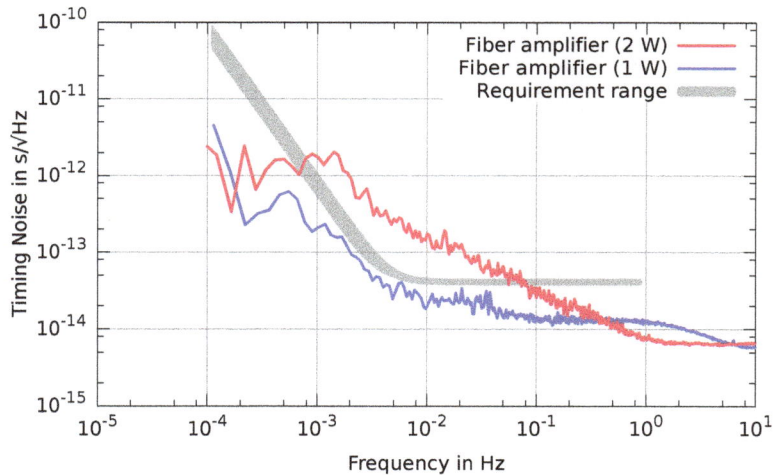

Figure 5.13: Timing jitter of an Ytterbium-doped fiber amplifier measured at 2 GHz modulation frequency for 1 and 2 W output power. The requirements are met for a power of 1 W but clearly violated for an output power of 2 W [66].

Ytterbium-doped fiber amplifier measured at 2 GHz modulation frequency for 1 and 2 W output power are taken from [123] and plotted as timing jitter in Figure 5.13. While at a power of 1 W, the timing jitter is low enough to meet the demanding level ten times below the carrier read-out noise, these requirements are clearly violated for an output power of 2 W.

The monolithic fiber amplifier used in the previously described relative intensity noise measurements ("PSFA-1064-01-10W-2-3" by *Nufern*, see Section 3.2.1) uses a large mode area fiber. The manufacturer claims that this technology suppresses stimulated Brillouin scattering and other non-linearities. Naturally, we hoped for an increased phase fidelity at higher output powers. While measurements on this device are still ongoing, preliminary results performed at a modulation frequency of 2.5 GHz are shown in Figure 5.14. It becomes obvious that—at least for an early testing unit provided

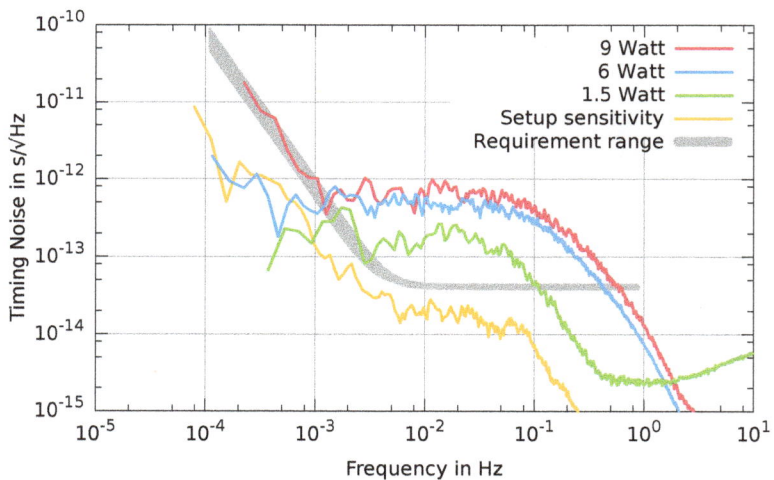

Figure 5.14: Timing jitter of the *Nufern* fiber amplifier measured at 2.5 GHz modulation frequency for 1.5, 6, and 9 W output power. The requirements cannot be met for either power level, a strong increase in noise can be observed with increasing power. The exact cause for the excess noise was not determined. The measurement sensitivity of the setup was sufficient.

by the manufacturer—phase noise increases significantly for higher powers. The timing stability requirements cannot be met for either power output although the sensitivity of the measurement setup was sufficient. Currently, phase fidelity measurements on a newer revision of this fiber amplifier are performed at the *Albert Einstein Institute* in Hanover, Germany.

The setup sensitivity can be determined with a simple optical fiber that replaces the fiber amplifier.

If might turn out that output powers greater than 1 W cannot be achieved within the stringent requirements. However, a lower total laser power can always be compensated with an increase in telescope diameter. Thus, in principle, a 1 W fiber amplifier would be sufficient to fly a successful mission.

Given the right choice of components and a sufficient temperature stability, **we could show that the excess timing jitter introduced by any component in the gigahertz signal path complies with the requirements for all discussed mission concepts**. Only phase noise introduced between carrier and sidebands by fiber amplifiers is an ongoing concern. It will be studied in detail as part of the development of a laser system for gravitational wave observatories.

5.2 FREQUENCY CONVERTERS

The pre-experiments on the GHz signal path set the boundary conditions and put requirements like the GHz signal's power on the frequency distribution system. We now have to identify components that meet the given requirements and examine the second signal path inside the FDS all the way to the analog-to-digital converters. This includes the frequency conversion between the GHz modulation signal and the MHz pilot tone signal.

Many different electronic devices had to be tested to assure that no excess timing jitter above the required level is produced by any of the components necessary for the conversion. Unlike the EOMs or fiber amplifiers, most active components that convert a signal cannot be tested individually. Wherever this is the case, we split the input signal and used two equal components. All uncorrelated noise between both devices then shows up in a differential measurement.

Differential noise between identical devices was reduced by a $\sqrt{2}$-factor to represent the noise of a single device throughout this chapter. **Keep in mind that all correlated noise will be subject to common mode noise suppression.**

To reveal correlated noise sources and nonlinearities, a '3-signal-test' that involves three different frequencies (and three identical chains of components) was proposed [132]. A hexagonal electro-optical test stand [133] that will allow for such tests is currently under construction at the *Albert Einstein Institute* in Hanover, Germany.

5.2.1 UP-CONVERSION VS. DOWN-CONVERSION

The frequency distribution system was developed to **Technology Readiness Level 4** specifications (component and/or breadboard validation in laboratory environment) within the scope of a contract between the *European Space Agency* (ESA) and *DTU Space* (National Space Institute, Denmark), the *Albert Einstein Institute* (Hanover, Germany), and *Axcon Aps* (Lyngby, Denmark).

The final goal was to develop a fully functional TRL 4 compliant electronic board that ensures a phase stable conversion between the GHz modulation signal and the MHz pilot tone signal (sine-wave). Additionally it must produce a differential MHz system clock as square-wave. Before we could start the development, a decision on the basic conversion principle had to be made. Only when the timing noise in the conversion process is below the required level, a sufficiently stable synchronization between all metrology systems within the constellation is possible.

One of both sine-wave signals can be produced directly by an on-board oscillator. You can either choose to convert the pilot tone up to the required GHz level, or convert a GHz modulation signal down to the pilot tone frequency. We first evaluated a fractional-N synthesizer that converted a 50 MHz signal up to 2 GHz. The results were later compared to the timing stability of a 2.016 GHz signal down-converted to 48 MHz (division by 42) by a programmable integer divider. Both conversion principles are illustrated in Figure 5.15. The synthesizer ("SYN2000ALC" by *Gronefeld*, Figure 5.16 left) is an all-in-one solution that can use the pilot tone as input and produces a modulation signal at the right level and frequency. The divider ("UXN14M9PE" by *Centellax*, Figure 5.16 right) uses a strongly attenuated modulation signal as input. It produces an unbalanced square wave (with a duty cycle smaller 50% and a DC offset) that we cannot use directly as pilot tone.

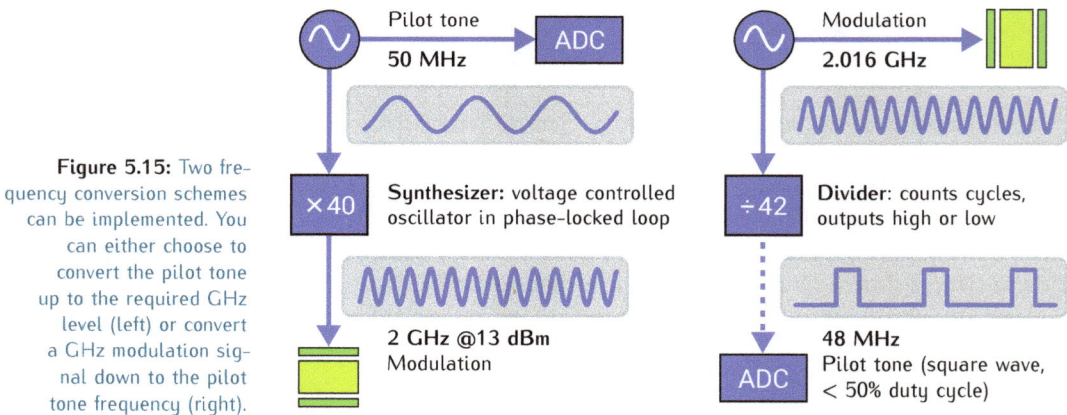

Figure 5.15: Two frequency conversion schemes can be implemented. You can either choose to convert the pilot tone up to the required GHz level (left) or convert a GHz modulation signal down to the pilot tone frequency (right).

Figure 5.16: *Gronefeld* SYN2000ALC fractional-N synthesizer (left) with BCD switches for output frequency adjustment and *Centellax* UXN14M9P programmable integer divider (right) on evaluation board.

5.2.2 FRACTIONAL-N SYNTHESIZER

The fractional-N synthesizers "SYN2000ALC" were custom made to our specifications by the *Ingenieurbüro Gronefeld*. They use an 8 GHz 16-Bit fractional-N PLL ("HMC700LP4" by *Hittite Microwave*) that locks an internal $f_{out} = 2000 \pm 20$ MHz voltage controlled oscillator ("ROS-2015+" by *Mini Circuits*) to a $f_{in} = 50$ MHz input signal at $0 \ldots 12$ dBm (sine or CMOS). The signal was amplified in two stages (two times "MGA-81563" by *Avago Technologies*) to 13 dBm output power. A space-qualified version of the same PLL chip is available. A precision voltage reference ("LM4130BIM-2.5" by *National Semiconductor*) was integrated to reduce close-to-carrier phase noise. The exact output frequency can be adjusted by BCD switches in 100 kHz steps.

The synthesizer can also operate with a **higher input frequency** after reprogramming the fractional-N PLL chip.

This device can be driven with a realistic pilot tone. The output is clean and powerful enough to drive an EOM directly. Only the system clock – that triggers the ADCs and needs to be at a frequency different from the pilot tone – would require additional electronics. Since no strict requirements apply to the system clock phase stability, the synthesizer covers all critical functionality of the necessary frequency converter.

We evaluated the phase fidelity between the input and output signal with an FPGA-based phasemeter capable of μcycle/$\sqrt{\text{Hz}}$ precision for frequencies up to 20 MHz [134, p. 94]. All measurements were performed at the University of Florida. In the test setup (see Figure 5.17) two synthesizers were driven by the same 50 MHz input signal. We set both devices to identical output frequencies. The outputs were individually mixed down with a common reference frequency to produce two 1 MHz signals. The phasemeter then measured the phase shifts in both signals. We concluded on the timing jitter introduced by any of the two synthesizers on basis of the differential phase noise between both devices.

To verify the basic functionality, we modulated the phase of one of the 50 MHz input signals with a slow sinusoidal shift of 2π mrad amplitude. The setup was modified to measure the signals' phase before and after up-conversion. As expected, the phase modulation is amplified in the up-converted

This means that timing noise is preserved by the synthesizer and the phase noise of the pilot tone is **amplified** as intended. Thus the device can be used to enhance the phase jitter signal in the sideband read-out as explained in Section 2.2.2.1.

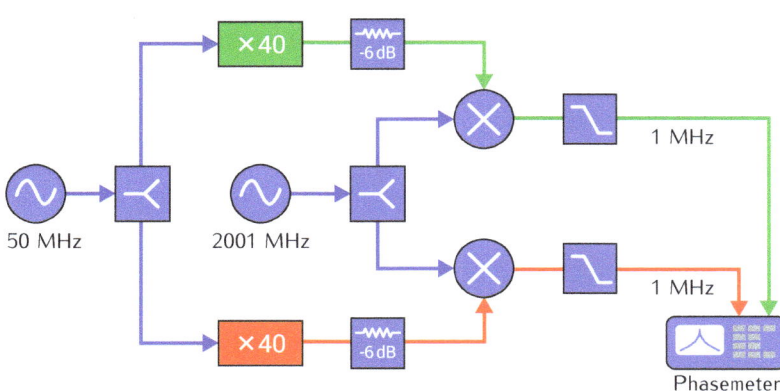

Figure 5.17: Two synthesizers driven by the same 50 MHz input signal are set to identical output frequencies. The outputs were individually mixed down with a common reference frequency to produce two 1 MHz signals.

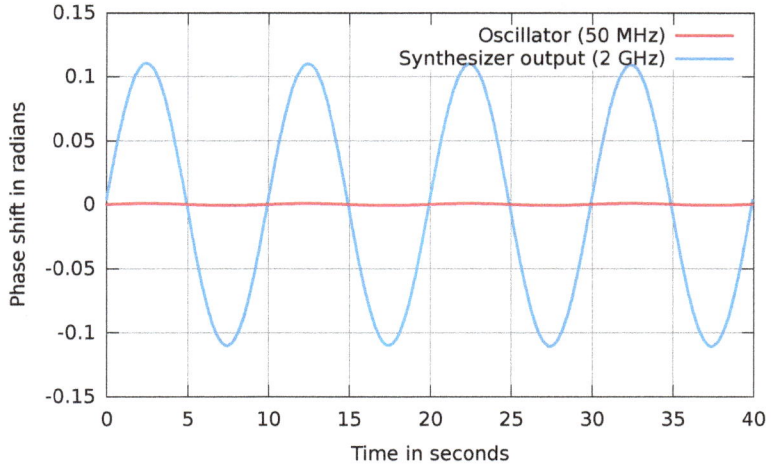

Figure 5.18: Phase shift of input and output signal for one *Gronefeld* "SYN2000ALC". The input signal (red) was provided by a sinusoidally phase modulated oscillator at 50 MHz. While the original phase modulation had a 2π mrad amplitude, the modulation was amplified in the frequency up-conversion (50 MHz to 2 GHz) by the same factor of 40 (blue).

2 GHz signal (see Section 4.3.2) by the factor $f_{out}/f_{in} = 40$. Results of this measurement are shown in Figure 5.18. While in this scale the phase modulation of the input signal is barely visible, the amplified modulation after up-conversion confirms the predicted behavior.

This value corresponds to 2.0×10^{-15} s/$\sqrt{\text{Hz}}$ at 100 Hz Fourier frequency and is well within the requirements.

The synthesizer was specified for a phase noise of -95 dBc/Hz at 100 Hz offset with regard to the 2 GHz output signal. The actual timing stability at the targeted measurement band was evaluated with the setup described above. Due to a careful selection of components (like splitters, attenuators, mixers, and low-pass filters) and a multi-layer thermal shield, the sensitivity of the setup is much better than required. Unfortunately, the synthesizers' timing jitter was significantly above the required level for almost the entire band as shown in Figure 5.19. The observed noise level was later verified independently by Ulrich Velte of the Institute for Quantum Optics (Leibniz Universität Hannover) [135].

The **setup sensitivity** was determined with a split 2 GHz signal that replaced the fractional-N synthesizers.

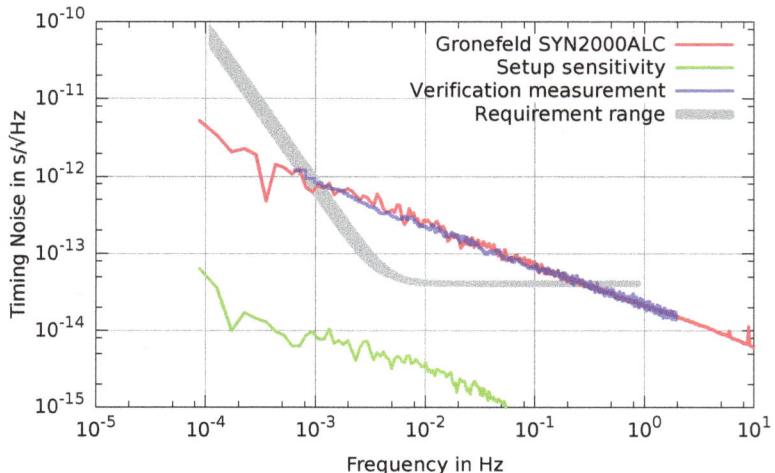

Figure 5.19: The timing jitter of the *Gronefeld* "SYN2000ALC" synthesizer (red) is significantly above the required level (gray) for frequencies between 1 mHz and 0.1 Hz. The setup sensitivity itself (green) was well within the required level. The observed noise level was verified independently by the Institute for Quantum Optics (Leibniz Universität Hannover) [135].

In an effort to lower the timing jitter, I extended the phasemeter's FPGA programming and associated read-out software to include the instantaneous amplitude (see Section 4.1) in the data stream. This allowed us to investigate the relationship between the power of the 50 MHz input and phase shifts in the synthesizer output. We varied the amplitude of the common input signal between 2.7 and 3.0 Vpp (12.6 . . . 13.5 dBm) and measured the phase of both output signals. Results are illustrated in Figure 5.20 and show a linear correlation.

From the above measurements, we derived a correlation between input amplitude and differential phase of 0.2 rad/Vpp. This commands an input signal stability of roughly 2.3×10^{-3} Vpp/$\sqrt{\text{Hz}}$ to fulfill the timing requirements. Measurements of the original input signal reveal an amplitude stability of better than 4.5×10^{-4} Vpp/$\sqrt{\text{Hz}}$. Thus the amplitude noise was not the limiting noise source in the initial measurement. As obvious from the absolute phase shifts in a single signal, most of the phase shifts due to amplitude fluctuations cancel via common mode noise suppression in the differential measurement. As stated earlier, such noise can easily be overseen in such a measurement scheme. For an actual gravitational wave observatory, oscillators that produce the input signal on the different spacecraft are of course independent and tougher requirements apply. When dealing with only one synthesizer the observed correlation of 4 rad/Vpp tightens the amplitude stability requirement to a much harder to reach value of 1.2×10^{-4} Vpp/$\sqrt{\text{Hz}}$.

Other phase correlations such as power supply voltage, temperature, or output amplitude were investigated, but none of them could be found to limit the phase fidelity performance of the synthesizers. The true cause of the observed timing jitter remains unknown and is probably related to intrinsic digital noise of the PLL chip itself, particularly by the inevitable frequency dividers embedded within.

The *Pentek* FPGA based phasemeter was originally designed to output the instantaneous frequency (PIR) only. I doubled the packing scheme (48 bits for PIR, 28 bits for I, 28 bits for Q data) to receive raw I and Q mixer outputs at 97.65 kHz and 23.84 Hz data rates for 2 or 4 channels. Due to limitations of the file transfer protocol, the PIR value could not be transfered simultaneously. "MATLAB", "Simulink", VHDL code, and the "ISE Design Suite 10.1" were used for **FPGA programming**.

We used a custom made 50 MHz TTL based **amplitude stabilization** based on a *Maxim* "MAX6126" ultra-low-noise voltage reference, see Section B.1 on page 169.

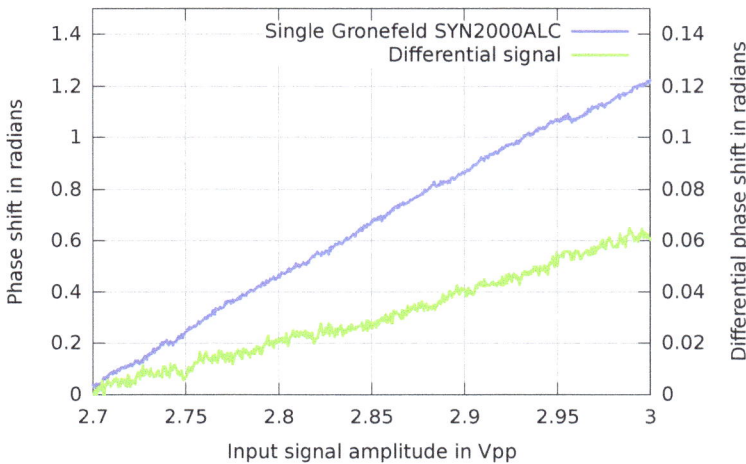

Figure 5.20: Correlation between input amplitude and differential phase for the *Gronefeld* "SYN2000ALC" synthesizer. A coefficient of 0.2 rad/Vpp is only valid for differential measurements (green, right hand y-axis). A single device features a much larger correlation of 4 rad/Vpp (blue, left-hand y-axis).

5.2.3 PROGRAMMABLE INTEGER DIVIDER

The "UXN14M9P" by *Centellax* is a programmable integer divider covering all integer divide ratios between 8 and 511 for maximum input frequencies of 14 GHz at $0 \ldots 10$ dBm input power. The output signal is specified as DC to 1.75 GHz at 4 dBm with an overall power consumption of ≈ 1.2 W (370 mA at 3.3 V). The device features single-ended or differential Current Mode Logic inputs and outputs. It is available as evaluation board *Centellax* "UXN14M9PE" that features 50 Ω differential inputs and outputs with SMA connectors. The component was chosen for its low SSB phase noise of -147 dBc at 10 kHz carrier offset specified at full division (/511) and related to 15.65 MHz (8 GHz input frequency).

For divide ratios between 16 and 511, the pulse width remains constant in each octave band. Thus the output is only **symmetrical** for powers of 2 and the duty cycle can be as low as 25% for other ratios. For divide ratios between 8 and 15 the duty cycle varies with the divide ratio, ranging from 33% to 64%.

As most integer dividers, the *Centellax* "UXN14M9P" produces a square wave output which is highly asymmetrical for almost all division ratios and additionally has a strong DC offset. For phase fidelity measurements, this offset was removed by bias-tees and the output was shaped to a sinusoidal wave by low-pass filters as shown in Figure 5.21. Since the expected phase noise is much smaller after frequency division, measurements are much more demanding on the phasemeter.

The overall setup to measure the timing stability of the divider is quasi identical to the one used for synthesizer evaluation (see Figure 5.17) but uses different frequencies. An input signal at 2016 MHz was divided by 42 and mixed down with a 48.0016 MHz reference to 1.6 kHz. Thus a software phasemeter could be used and the differential phase noise between both channels was converted to timing jitter. The measured timing jitter shown in Figure 5.22 (red trace) is marginally below the required level.

To verify the positive results, a second measurement setup was designed that mixes the output of both dividers directly to DC (see Figure 5.23). We tuned the setup to a constant output of 0 V so that phase shifts by either divider would result in a linear voltage increase or decrease of the mixer output. A simple data acquisition card was used to record the voltage shifts over time. The scaling factor between DC value and phase shift was determined. Voltage noise projected to timing jitter is shown in Figure 5.22 (blue trace). The result of this measurement even slightly improved upon the differential

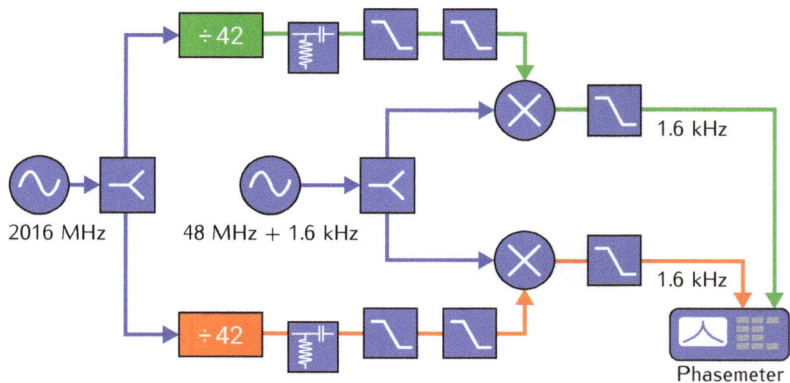

Figure 5.21: An input signal at 2016 MHz was split and divided by 42. Offsets were removed by bias-tees and the output was shaped to a sinusoidal wave by low-pass filters. The signal was mixed down with a 48.0016 MHz reference to 1.6 kHz.

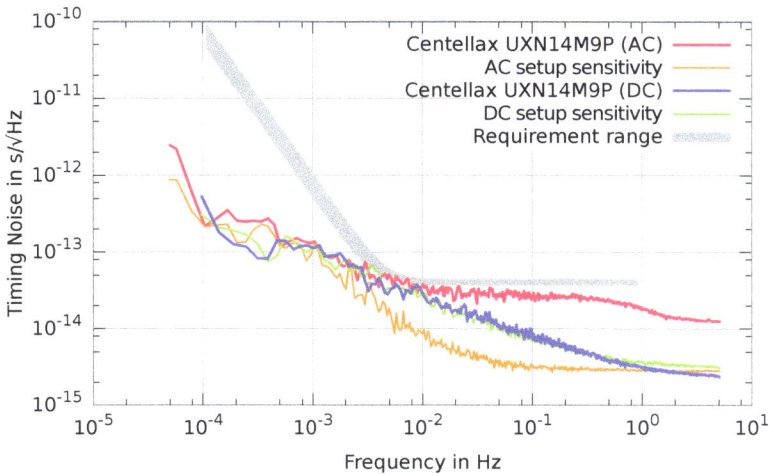

Figure 5.22: The differential noise between two *Centellax* "UXN14M9P" integer dividers (red, blue) meets the required level (gray) over almost the entire measurement range. The phase noise introduced by the setup itself (green, orange) seem to cause the marginal excess noise at around 5 mHz for the AC measurement.

phase noise making the *Centellax* "UXN14M9PE" a viable candidate for the Frequency Distribution System.

The sensitivity of the setup, is easily being degraded by a number of effects such as the actual signal level available for the frequency mixers. This might explain the excess noise present in the AC measurement.

The **setup sensitivity** was measured with one divider output signal split for two bias-tees.

Considering that the divider measurements were early results and no extraordinary measures for temperature stabilization were implemented, we were confident that the performance of the device would improve given the right conditions. Since even a fractional-N synthesizer built specifically to our needs showed excess phase noise way above the requirements, **all future design effort for the Frequency Distribution System followed a phase noise down-conversion scheme with digital dividers**. Although the divider output cannot be used as pilot tone directly and further signal shaping is necessary, dividers are much simpler in nature and easier to handle than the more complex multipliers.

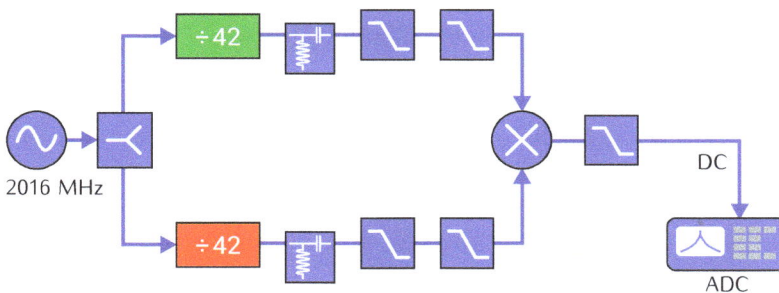

Figure 5.23: Both dividers are mixed directly to DC. Phase shifts by either divider result in a linear voltage increase or decrease of the mixer output that can be projected to phase noise.

5.3 MEGAHERTZ SIGNAL PATH

The following research was part of a technology development activity by the *European Space Agency* conducted between June 2011 and February 2014 under the contract number AO/1-6238/10/NL/HB. The LISA Metrology System Team – with members from *DTU Space* (National Space Institute, Denmark), the *Albert Einstein Institute* (Hanover, Germany), and *Axcon Aps* (Lyngby, Denmark) – compiled a number of technical notes [115, 136–148] that hold further details. Availability can be inquired through the *European Space Research and Technology Centre*. Additionally, a publicly available final report [74] features a good overview of the overall scope of the activity.

As a member of the LISA Metrology System Team I was primarily responsible for the design and testing of the Frequency Distribution System.

The decision to drop the up-conversion scheme and concentrate on digital dividers was made at the very beginning of the project. The potentially highly asymmetrical square-wave output of digital dividers creates new challenges though. While the system clock is expected to be a differential MHz square wave signal, the pilot tone has to be a low-distortion MHz sine wave. Thus an additional filter in combination with amplifiers and power splitters for signal distribution is necessary when pursuing a division scheme for the final Frequency Distribution System. The components are different for the pilot tone and system clock generation as shown in Figure 5.24. The planned chain of components for the clock generation are:

❶ Integer divider (to divide from GHz modulation signal down to twice the pilot tone / system clock frequency)

❷ Digital prescaler (divide-by-2 flipflop to get signal with 50% duty cycle)

For the pilot tone generation there are the following additional components:

❸ Buffer amplifier (to boost and stabilize the amplitude)

❹ Shaping filter (5th order, to filter out higher harmonics)

❺ Power splitters (to distribute the pilot tone to the different ADCs)

❻ Cable bridges with high-pass filters (to pass the pilot tone to the individual ADC boards and reduce crosstalk)

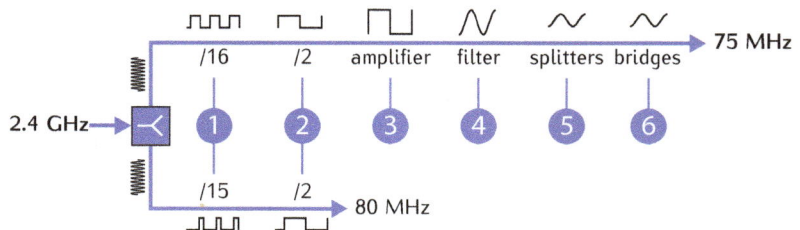

Figure 5.24: Planned chain of components for pilot tone and system clock generation and distribution.

Figure 5.25: One component chain (prescaler, buffer amplifier, shaping filter, power splitters) of the FDS evaluation prototyping board.

Additionally, the GHz power splitter and mixer as well as the filters and the power combiner for the PRN code are necessary. All these components have to handle signals on the order of roughly 50 to 200 MHz and thus lie in the megahertz signal path. Figure 5.24 already uses the latest frequencies and divide ratios as explained in Section 4.4, including an 80 MHz system clock and a 75 MHz pilot tone. These frequencies are our current baseline.

Early on within the scope of this project, *Axcon ApS* designed two identical prototype Frequency Distribution System boards that covered components ❷ to ❺: Both boards featured *Peregrine Semiconductor* "PE3511" digital prescalers, *ON Semiconductor* "NB3L553" buffer amplifiers, custom-designed 5th order shaping filters by *Axcon Aps*, and a chain of *Mini Circuits* "ADP-2-1W+" power splitters. Figure 5.25 shows one of these component chains (see Figure 5.27.5 for a picture of the full system). The specific components were chosen based on educated guesses and were subject to performance evaluation.

The measurement principle is almost identical to the one shown in Figure 5.21. A 152 MHz input signal was provided by a *Centellax* "UXN14M9P" evaluation board and split for both prototype boards. This procedure delivers realistic signals while noise of the Centellax divider itself was common mode and thus suppressed in the differential phase noise measurement. Results converted to timing jitter in the 76 MHz output signal are shown in Figure 5.26. It is obvious that we had to evaluate the components individually and find alternatives for those that violate the timing stability requirements. Many different components were tested, some of which are shown in Figure 5.27. I will now present a selection of the most important results.

Additional **power splitters and mixers** are required for the differential phase noise read-out between the two GHz references, see Section 4.4.
Filters and power combiner were already tested successfully, see Section 5.1.2.1.
The **baseline** set of frequencies was adjusted multiple times in the past and some tests were done with slightly different frequencies.

While the basic principle is **identical**, each device brings its own challenges such as different requirements for temperature or supply voltage stability. Thus the detailed setup has to be adapted.

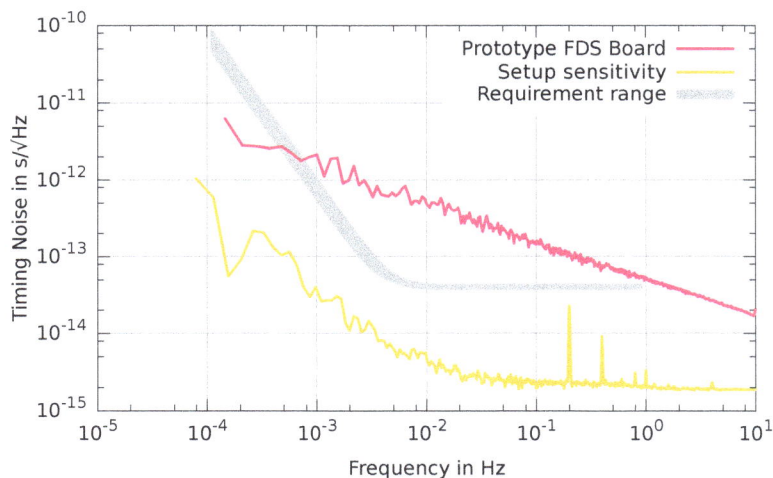

Figure 5.26: Differential timing stability between two prototype FDS boards measured at 76 MHz. A 152 MHz square-wave input signal was provided by a *Centellax* "UXN14M9P" evaluation board and split for both prototype boards.

Figure 5.27: Selection of components evaluated for the Frequency Distribution System board.

Figure 5.27.1: *ON Semiconductor* "NB7L32MMNEVB" prescaler evaluation board.

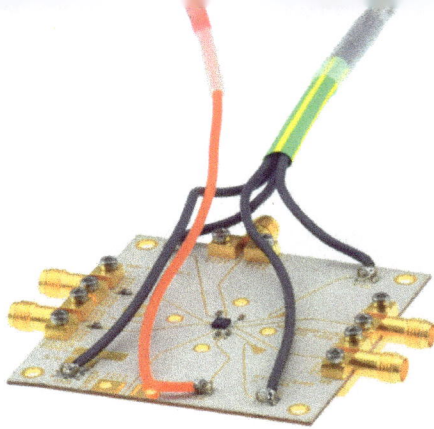

Figure 5.27.2: Multiple *Peregrine Semiconductor* divider evaluation boards, some were radiation hard and compatible with space applications.

Figure 5.27.3: *ON Semiconductor* "NB6N239SMNEVB" prescaler evaluation board.

Figure 5.27.4: *Mini Circuits* GHz splitter with mixer and low-pass filter.

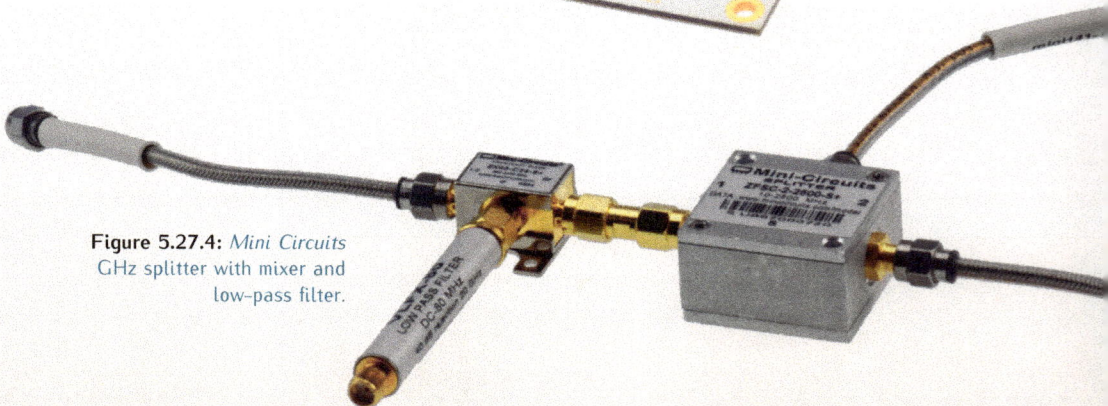

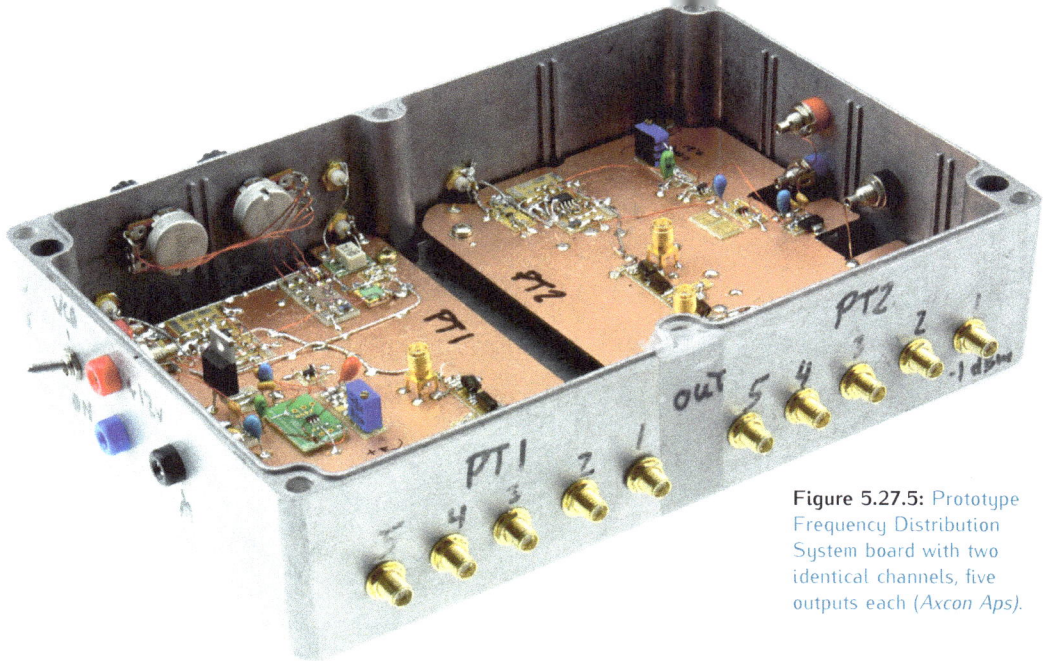

Figure 5.27.5: Prototype Frequency Distribution System board with two identical channels, five outputs each (*Axcon Aps*).

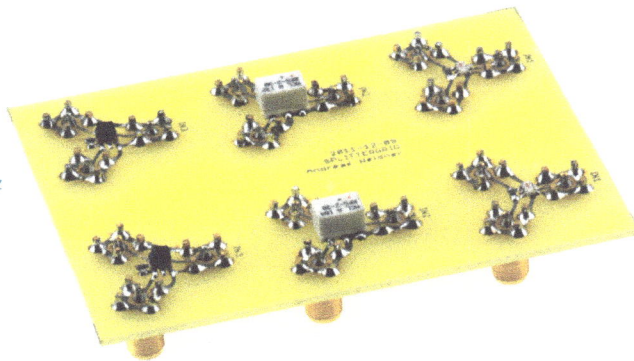

Figure 5.27.6: Custom GHz power splitters evaluation board with three different devices, two channels each (*Albert Einstein Institute*).

Figure 5.27.7: Cable bridge ("PhaseTrack210" by *Times Microwave Systems*) with two high-pass filters.

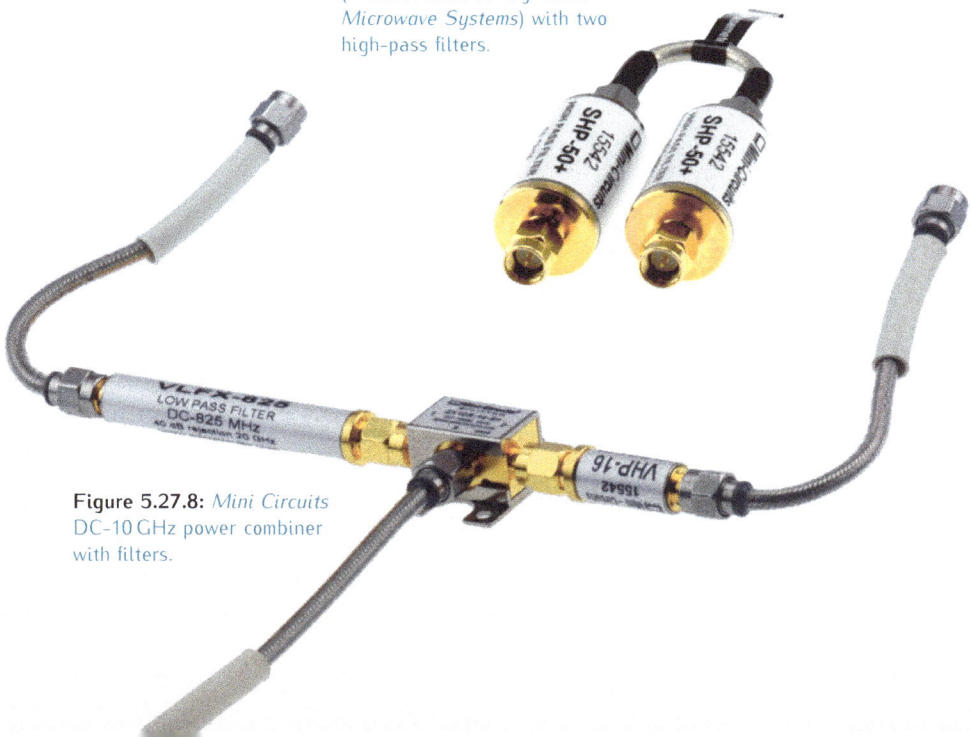

Figure 5.27.8: *Mini Circuits* DC-10 GHz power combiner with filters.

5.3.1 INTEGER DIVIDER AND DIGITAL PRESCALERS

We tested a number of different integer dividers and digital prescalers as alternatives to the *Centellax* "UXN14M9P". Some *Peregrine Semiconductor* dividers (see Figure 5.27.2) were even radiation hard and generally compatible with space applications. A selection of dividers for which measurements are presented in this chapter are listed in Table 11 together with the most important specifications for these dividers.

Table 11: Selection of dividers that were tested within the scope of the "LISA Metrology System" development.

*Compatible with space applications

Digital divider	Input frequency	Divide ratio
Peregrine Semiconductor "PE9303"*	1.5...3.5 GHz	8
Peregrine Semiconductor "PE3513"	DC...1.5 GHz	8
Peregrine Semiconductor "PE3511"	DC...1.5 GHz	2
Centellax "UXN14M9PE"	DC...14 GHz	8...511
ON Semiconductor "NB6N239SMNEVB"	DC...3.0 GHz	2/4/8/16
ON Semiconductor "NB7L32MMNEVB"	DC...14 GHz	2

The *ON Semiconductor* "NB6N239SMN" (see Figure 5.27.3) offered a selectable divide ratio, but all at **powers of two**.

The custom **board** was made of a PCB material with very similar properties as the one used for the manufacturer provided evaluation board.

All dividers except for the one by *Centellax* had a divide ratio at powers of two and hence limited the choice of pilot tone frequencies. Results for integer divider candidates (see ❶) or alternative components used in a chain of digital prescalers are presented in Figure 5.28. Most differential phase noise measurements were performed between two manufacturer provided evaluation boards for a variety of different frequencies. A chain of five *ON Semiconductor* "NB7L32MMN" dividers (see Figure 5.27.1) with a divide ratio of 2 was evaluated on a board designed and manufactured by *Axcon Aps*. This divider chain shows a timing jitter about five times larger than for a single device. However, in a perfect world and with a constant timing jitter over frequency, the theoretical factor should be $\sqrt{5}$ for five uncorrelated noise sources. The discrepancy can be explained by an increase of timing jitter at lower frequen-

Figure 5.28: Timing stability of different integer dividers and digital prescalers. *Peregrine Semiconductor* "PE9303" for 2.4 GHz /8, *Peregrine Semiconductor* "PE3513" for 600 MHz /8, *ON Semiconductor* "NB7L32MNMNEVB" for 2.4 GHz /2 (chain of five "NB7L32MMN" for 2.4 GHz /32), *Centellax* "UXN14M9PE" for 2016 MHz /42.

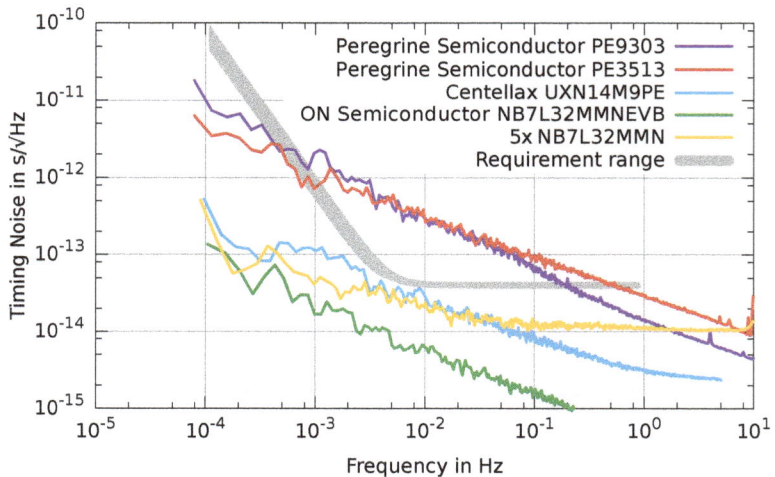

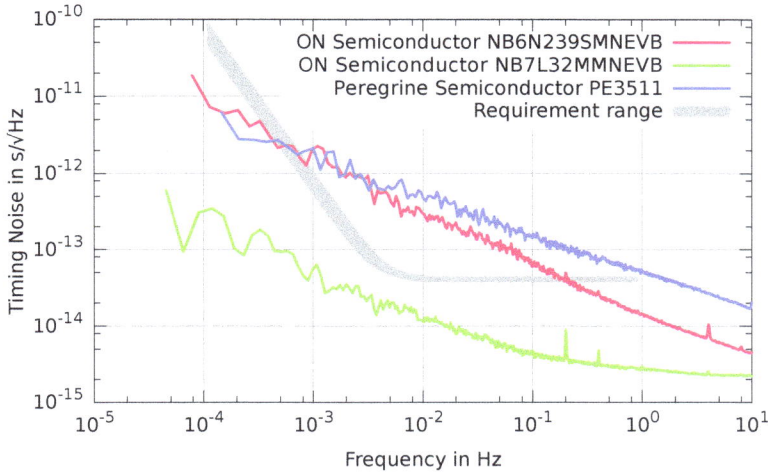

Figure 5.29: Timing stability of different divide-by-two digital prescalers. *ON Semiconductor "NB6N239SMNEVB" for 150 MHz /2, ON Semiconductor "NB7L32MNMNEVB" for 150 MHz /2, Peregrine Semiconductor "PE3511" for 152 MHz /2.*

cies. The measurement presented in Figure 5.28 was performed at 2.4 GHz. For comparison, a measurement for a 150 MHz signal is presented in Figure 5.29 that shows a timing jitter which is larger by a factor of 2 to 3.

To have a wider range of possible output frequencies, an integer divider like the *Centellax* "UXN14M9P" is necessary. For the current baseline set of frequencies, this is at least true for the system clock generation (2.4 GHz to 80 MHz) which cannot be achieved by a divide ratio of any power of two. Unfortunately, the *Centellax* divider provides a highly asymmetric signal (27% duty cycle) at the designated divide ratio of 30. This issue can be avoided by a divide ratio of 15 and a subsequent prescaler stage (see ❷) that generates a signal with 50% duty cycle at half the frequency. Prescaler candidates with a divide ratio of 2 measured at relevant frequencies are presented in Figure 5.29. As it turned out, the *Peregrine Semiconductor* "PE3511" digital prescaler was the dominant noise source in the prototype Frequency Distribution System board.

5.3.2 BUFFER AMPLIFIER

An *ON Semiconductor* "NB3L553" 1:4 clock fanout buffer was used as buffer amplifier (see ❸) in the prototype board. This device was another limiting noise source. An amplifier is required since the output signal of most digital dividers is not powerful enough to act as pilot tone for the many channels of the Phase Measurement System. An alternative device ("RAMP-33LN" by *Mini Circuits*) was evaluated and found to comply with the timing stability requirements. The "RAMP-33LN" is also available as coaxial version in a shielded housing (*Mini Circuits* "ZX60-33LN+'). We tested all devices differentially with a square-wave input signal at a representative signal amplitude and for relevant frequencies. Results are shown in Figure 5.30.

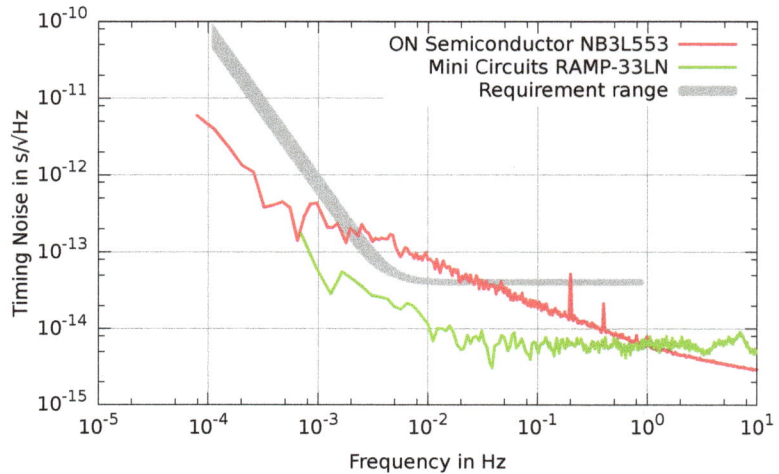

Figure 5.30: Timing stability of different buffer amplifiers. *ON Semiconductor* "NB3L553" measured at 76 MHz, *Mini Circuits* "RAMP-33LN" at 75 MHz, both with square-wave input signals.

5.3.3 FILTERS, POWER SPLITTERS, AND CABLE BRIDGES

A custom 5th order shaping filter (see ❹) designed by *Axcon Aps* converts the square-wave divider output to a sinusoidal signal. A chain of *Mini Circuits* "ADP-2-1W+" power splitters (see ❺) then distributes the pilot tone to as many as eight outputs. These two components of the prototype Frequency Distribution System board turned out to meet the required timing stability as is. Figure 5.31 shows the timing jitter of the shaping filter and the entire power splitter chain. No alternatives were evaluated.

Cable bridges to connect the final FDS with the individual ADC cards (see ❻) were designed using the best cable known to us ("PhaseTrack210" by *Times Microwave Systems*, see Section 5.1.1) in a semi rigid version. These cables were custom-made applying the minimum allowed bend radius to provide a connection as short as possible (see Figure 5.27.7).

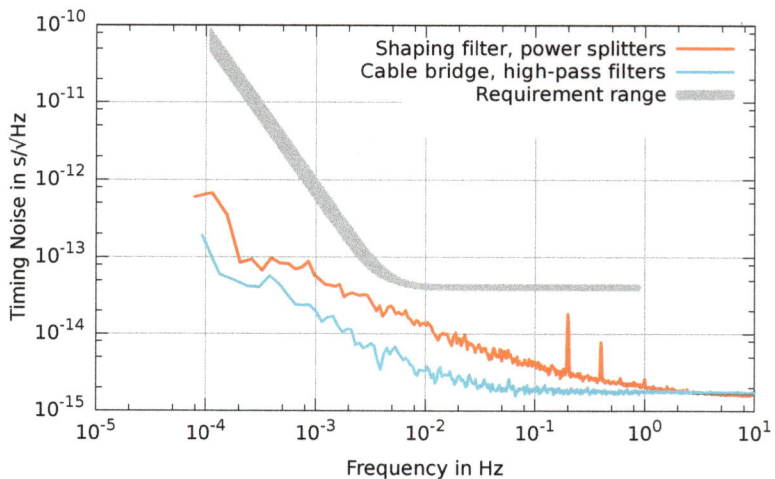

Figure 5.31: Timing stability of custom shaping filter and *Mini Circuits* "ADP-2-1W+" power splitters (measured at 76 MHz) and *Times Microwave Systems* "PhaseTrack210" semi-regid cable bridge with two *Mini Circuits* "SHP-50" high-pass filters (measured at 75 MHz).

To suppress crosstalk between individual ADC cards, we installed additional high-pass filters ("SHP-50" by *Mini Circuits*, two per cable bridge). The superb phase fidelity of these components is currently limited by the phasemeter sensitivity and probably even better than plotted in Figure 5.31.

In our case, **crosstalk** for heterodyne beat notes – that are added to the pilot tone on the ADC cards – are reduced by > 40 dB below the pass band of 41 ... 800 MHz before they can enter the FDS.

5.3.4 GHZ POWER SPLITTERS, MIXER, AND PRN CODE COMBINER

To determine the differential phase noise between two GHz reference oscillators, both GHz frequencies are mixed down to 1 MHz and measured by the phasemeter. This requires power splitters, UHF mixers, and low-pass filters. A total of three surface-mount GHz power splitters by *Mini Circuits* (see Figure 5.27.6) were evaluated. Although not specified for 2.4 GHz, the "SCN-2-22" showed the best performance and features the lowest loss as shown in Figure 5.32. Additionally, a well-shielded coaxial power splitter (*Mini Circuits* "ZFSC-2-2500-S") was evaluated (see Figure 5.27.4). The timing stability is plotted in Figure 5.32 in combination with a Level 7 frequency mixer (*Mini Circuits* "ZX05-C24-S", 300 ... 2400 MHz) and a 40 dB low-pass filter (*Mini Circuits* "VLFX-80", pass band DC... 80 MHz). Table 12 lists the specifications for the different splitters.

Power splitter	Frequency range	Loss at 2.4 GHz
Mini Circuits "SCN-2-22"	1850...2200 MHz	3.37...3.59 dB
Mini Circuits "RPS-2-30"	10...3000 MHz	3.61...4.24 dB
Mini Circuits "TCP-2-33W"	50...3000 MHz	3.42...3.95 dB
Mini Circuits "TCP-2-33W"*	10...2500 MHz	3.33...3.36 dB

Table 12: Different UHF power splitters tested to split the GHz modulation signal between pilot tone generation divider chain and reference oscillator mixer.

*Coaxial version

The GHz phase stability of the power combiner (*Mini Circuits* "ZX10R-14+") with associated filters (*Mini Circuits* "VHP-16" and "VLFX-825') is required for a simultaneous modulation of sidebands and PRN code (see Figure 5.27.8). Measurement results were already presented in Section 5.1.2.1. It was verified that the components and sidebands present in the DS/SS modulation do not affect the ranging accuracy [75].

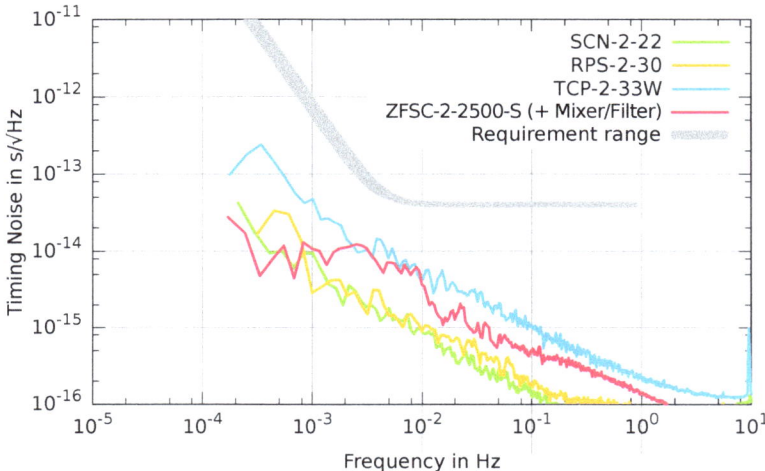

Figure 5.32: Timing stability of different UHF power splitters, mixers and filters. All components by *Mini Circuits*. Measurements performed at 2.5 GHz for the surface mount power splitters "SCN-2-22", "RPS-2-30" and "TCP-2-33W". The coaxial power splitter "ZFSC-2-2500-S" was tested in combination with frequency mixer "ZX05-C24-S" and low-pass filter "VLFX-80' at 2.4 GHz.

5.4 FULL SYSTEM EVALUATION

Based on the extensive market research and previous experiments, components were selected to build different versions of the Frequency Distribution System compliant with Technology Readiness Level 4. Two revisions of the PCB boards are shown in Figure 5.38. The system was designed to fit into the modular structure of the "LISA Metrology System" developed under ESA contract. Figure 5.42 shows the entire setup with the FDS connected to five ADC cards.

5.4.1 REVISION 1

For a first version of the TRL 4 FDS board, we chose the exact same principle to generate both, the pilot tone and the system clock: one integer divider (*Centellax* "UXN14M9P") set to a divide ratio of 16 (pilot tone) or 15 (system clock) followed by a single divide-by-two stage (*ON Semiconductor* "NB7L32MMN"). This generates a differential 80 MHz system clock and a 75 MHz square wave of 50% duty cycle. The pilot tone is amplified by a *Mini Circuits* "RAMP-33LN" with a subsequent attenuation stage to adjust the signal level. We integrated two chains of dividers and amplifiers for the 75 MHz signal alongside optimized versions of the 5th order shaping filter and power splitter chain taken from the prototype board. Hence we were able to perform differential phase noise measurements between two pilot tone generation chains on one single FDS board. Multiple SMA test points made it possible to evaluate smaller sections of the chain individually.

The board layout is shown in Figure 5.33 . The two divider stages for the pilot tone generation are featured in a magnified area. The shaping filters are arranged in parallel, right next to the on-board power supply. The pilot tone distribution with is located in a large area (top right), the system clock is generated at the top left area. A full block diagram can be found in Section B.2 on page 170.

Figure 5.33: Frequency Distribution Board board (revision 1) – Starting at the lower left corner, inputs J1 and J2 are connected to 2.400 GHz and 2.401 GHz respectively. Output J6 provides the mixed-down and low-pass filtered differential phase noise between both GHz signals. U3/U10 and U4/U11 are two equal pilot tone divider chains with subsequent amplifiers, attenuators, and shaping filters. The U3/U10 chain is connected to the pilot tone distribution on the upper right hand side. The clock signal is generated by dividers U5/U6 (top), the power supply for all dividers and amplifiers is located on the same board (left hand side). The magnified section (lower right) shows the two pilot tone divider chains in detail.

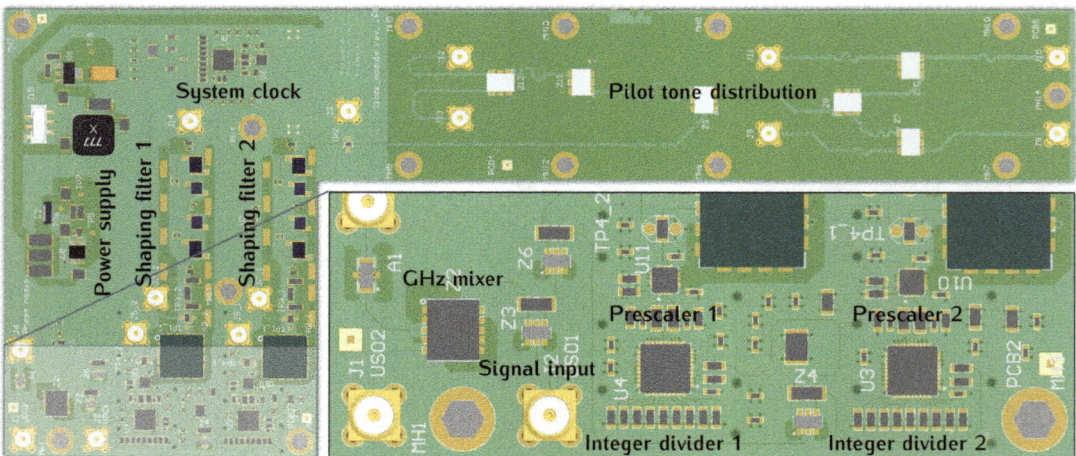

5.4.1.1 TEMPERATURE AS A DESIGN DRIVER

To avoid vias, we used only the top layer of the board for the pilot tone signals throughout the entire generation and distribution. Each individual signal line in the pilot tone distribution chain had to be as identical as possible. *Axcon Aps* placed all *Mini Circuits* "ADP-2-1W+" power splitters symmetrically and matched all traces in length. Critical traces are even matched in x and y direction considering the fact that the PCB may not expand homogeneously in all dimensions. This leads to the complex shape of traces shown in Figure 5.34.

Such an inhomogeneous expansion would lead to a relative phase shift between individual pilot tones on one FDS board.

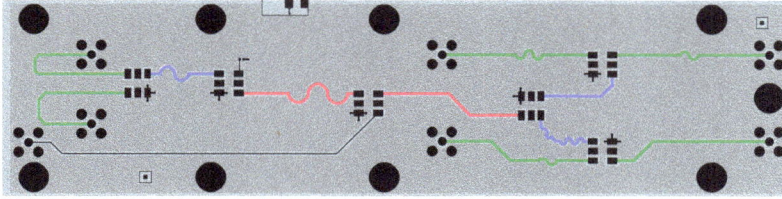

Figure 5.34: Pilot tone distribution PCB trace layer. Traces of individual colors are matched in length to each other.

The 5th order band pass was designed by *Axcon Aps* for a high attenuation of pilot tone harmonics. Due to the square wave characteristics, especially the 3rd harmonic (225 MHz) and 5th harmonic (375 MHz) frequencies were of concern. Simulations predicted a signal level of −83 dB below the pilot tone for the 3rd harmonic (see Figure 5.35). This translated to a still very good value of −75 dB in the final hardware.

Harmonic frequencies of the pilot tone may have to be considered in the frequency plan as forbidden frequencies if not attenuated sufficiently, see Section 3.2.

Additionally, phase shifts in the pass band due to component instabilities had to be addressed. To compensate a positive phase shift over temperature coefficient in coils used for the filter, one of the capacitors was especially chosen to have a compensating negative coefficient. All other capacitors were selected to have nominal zero phase shift over temperature change. With all components at their worst specified stabilities, a maximum of 7.8×10^{-4} rad/K at 75 MHz is expected. The phase over frequency change coefficient was designed to be 1.4×10^{-4} rad/Hz.

The **exact values** of all components of the shaping filter can be found in [115].

At **75 MHz** this phase shift corresponds to a change in signal arrival time per Kelvin of 1.7×10^{-12} s/K.

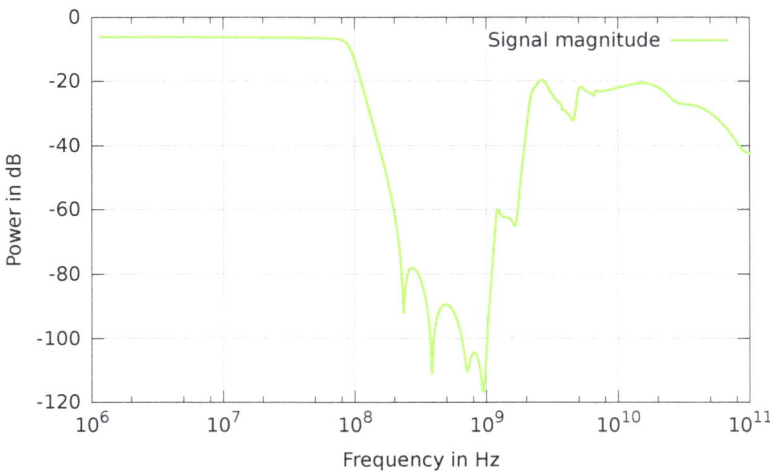

Figure 5.35: Simulated transfer function of 5th order shaping filter which was designed to have the strongest attenuation at uneven harmonics of the pilot tone. At 225 MHz, −83 dB with respect to the pass band were predicted. A value of −75 dB was achieved in real hardware.

5.4.1.2 PERFORMANCE MEASUREMENTS

We performed differential phase fidelity measurements between pilot tone divider stages, amplifiers, and shaping filters on one board. Using the pilot tone distribution for only one of the signals additionally results in a measurement of the absolute noise introduced by the signal lines and power splitters. The timing stability of the shaping filters and pilot distribution is excellent and partly limited by the setup sensitivity (see Figure 5.36). The combination of *Centellax* "UXN14M9P" and *ON Semiconducor* "NB7L32MMN" however generated a timing jitter significantly above the required level (red trace).

For comparison, the exact same combination of integer divider and prescaler is shown as yellow trace, this time measured between two chains of original evaluation boards. Prior to this measurement we assumed the more compact design to be more stable than a combination of evaluation boards. This is obviously not the case.

To rule out any other causes for excess phase noise, the on-board power splitters were removed and replaced by external shielded power splitters, which had been separately verified to be within requirements and which are commonly used in our laboratories for these kind of experiments. The observed noise remained the same though and could not be reduced by any available measures. Thus this excess noise was in fact generated by the divider chain. The root cause had to lie in the different implementation of individual components between the revision 1 FDS board and the corresponding divider evaluation boards.

For example, the power supply for the *Centellax* "UXN14M9P" dividers was reversed in accordance with its data sheet to reduce the number of signal levels shifts in the system from negative to positive logic. Additionally, a

Figure 5.36: Differential phase noise measurements at 75 MHz for the 1st revision FDS board. The timing stability of the shaping filters and pilot distribution (blue) is excellent and partly limited by the setup sensitivity (green). The divider chain on the FDS board that generates the pilot tone shows excess noise (red) while a combination of the same devices on the original evaluation boards (yellow) complies with the requirements.

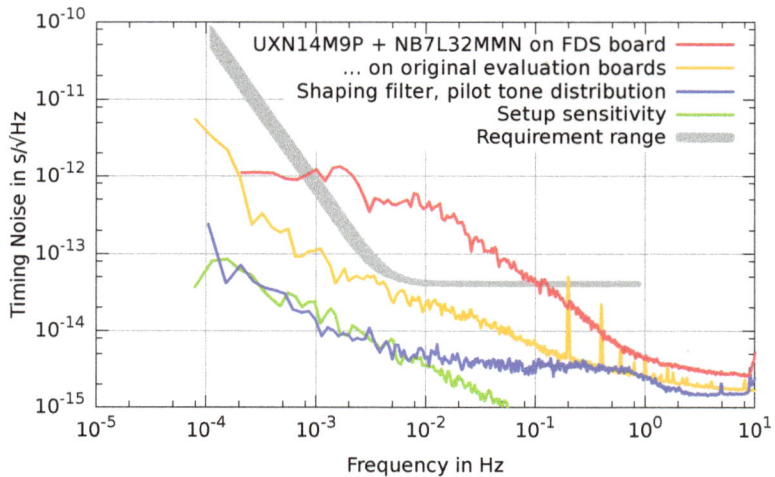

5.37.1: Visible spectrum.

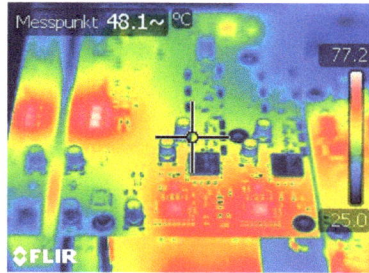

5.37.2: Infrared spectrum.

Figure 5.37: Frequency Distribution System board (revision 1) in visible (left) and infrared (right) spectrum. One can clearly see a strong temperature gradient across the shaping filters that originates at the power supply (left hand side) and the dividers (bottom). The color scale (right) is given in °C.

different PCB material was used: the evaluation board used *Rogers* "RO6002" which is very hard to come by in Europe.

As it turned out, the current board entraps heat which was made visible by a thermographic camera. Figure 5.37 (right) shows high temperatures in the area of the on-board power supply and the dividers. Temperatures peak at over 75°C. On top of that, a strong temperature gradient is visible across the PCB board, falling off diagonally along the shaping filters. The dividers on the FDS board are placed in close proximity to each other and become substantially hotter compared to the evaluation boards. One reason for that could be the thermal connection of the *Centellax* dividers which is at bottom of these devices and could not be reached in the current PCB layout. This made it very difficult to get the heat off the board.

Any of these differences in implementation could in theory account for the surprisingly high level of observed excess noise. Additionally, electromagnetic interference between the dividers could in principle add timing jitter. Effective shielding was hard to implement on the current revision so that the PCB layout was revised. This gave us the opportunity to address all of the above mentioned issues at once within the same revision effort.

5.4.2 REVISION 2

From what we had learned we designed a second revision of the FDS board. We kept all components that passed the tests, removed everything which was not essential and might cause excess noise or act as additional heat source, and improved some details. As part of this revision process, we decided to substitute the *Centellax* "UXN14M9P" in the pilot tone generation chain by five additional *ON Semiconductor* "NB7L32MMN" dividers. Although this limits possible divide ratios to powers of two, a chain of these prescalers was found to surpass the performance of the single integer divider (see Section 5.3.1).

Figure 5.38 shows revisions 1 and 2 of the Frequency Distribution System board side by side for comparison.

Figure 5.38: Frequency Distribution System boards, revisions 1 and 2. Layout by Axcon Aps.

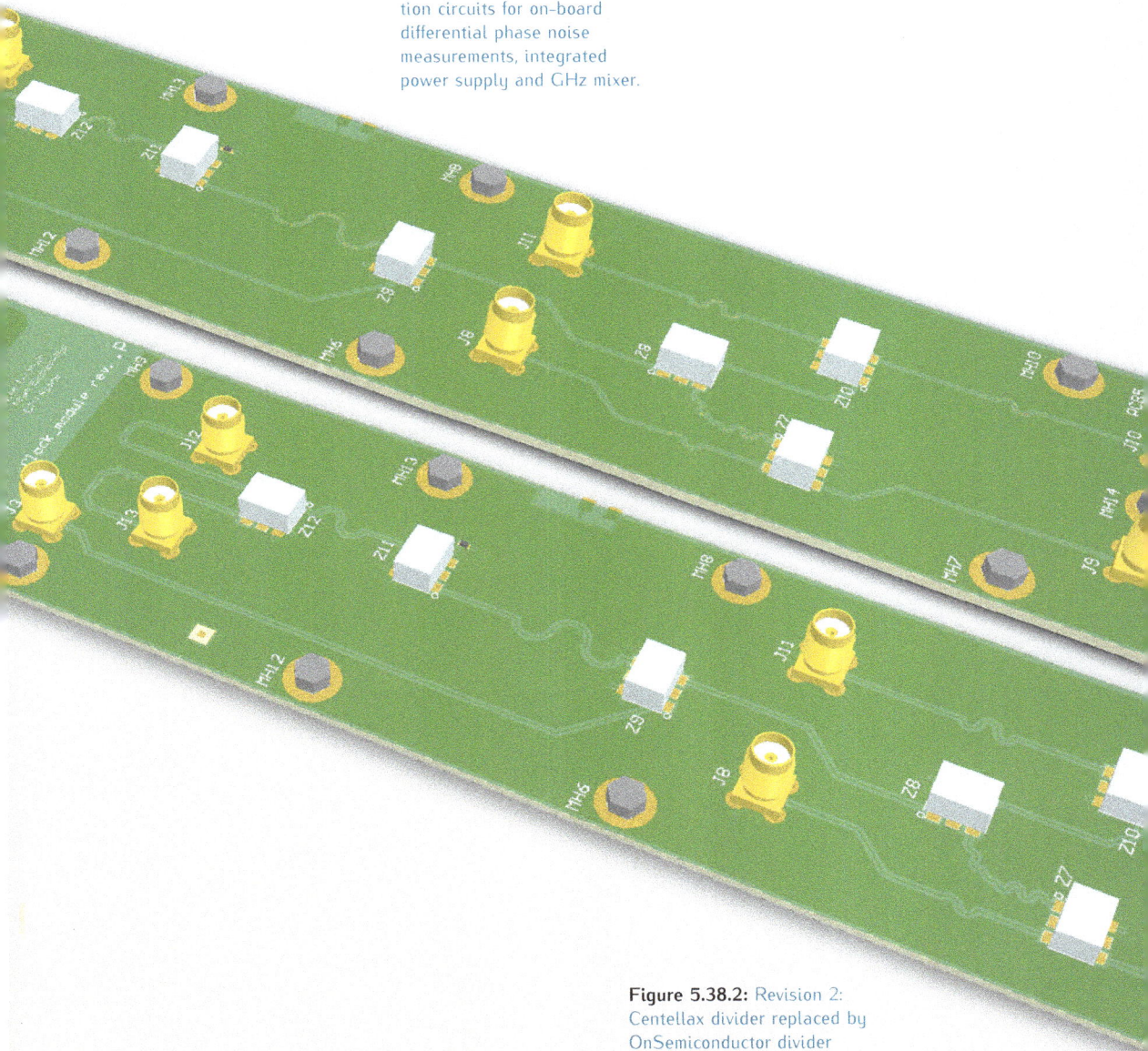

Figure 5.38.1: Revision 1: 2 identical pilot tone generation circuits for on-board differential phase noise measurements, integrated power supply and GHz mixer.

Figure 5.38.2: Revision 2: Centellax divider replaced by OnSemiconductor divider chain. 2nd pilot tone generation, on-board power supply, and GHz mixer removed.

Here is what we did to arrive at a second revision FDS board:

KEEP   We kept the shaping filters, pilot tone distribution power splitters as well as the high-pass filters and cable bridges from the revision 1 FDS board. Additionally, we reused the current divider chain of the *Centellax* "UXN14M9P" and *ON Semiconductor* "NB7L32MMN" for the system clock generation since the phase stability requirements are relaxed for this signal and the current solution is sufficient here.

REMOVE   We removed as many components as possible off the board to reduce any possible thermal or electromagnetic disturbances. This included the mixer to generate a correction signal for a second GHz signal as well as all GHz power splitters. Both components were originally integrated in the first revision but actually never used. A stabilized supply voltage is provided separately and the second revision FDS board now offers connectors to power different component groups individually.

IMPROVE   We implemented separate ground planes for different parts of the board and the amplifiers have a separate power supply plane which can be disabled. All components now have solder points for electric shielding, and thermal pads are connected through the backside of the board for better heat distribution and dissipation. To ensure the same performance as provided by the evaluation boards, we built the second revision FDS board with a primary trace layer of *Nelco* "N4000-13 EP" material.

The *Nelco* material performs very similar to the *Rogers* "RO6002" used in the *ON Semiconductor* evaluation board but is readily available in Europe.

REDESIGN   Since the *ON Semiconductor* "NB7L32MMN" digital prescaler outperforms all other dividers, we used these in a cascade to replace the hot and marginally performing *Centellax* "UXN14M9P". There are now five stages of digital prescalers to reach a divide ratio of 32, and an optional additional stage—that will be bypassed and powered off by default—to divide by a total ratio of 64. This additional stage adds flexibility in the pilot tone frequency and enables us to conduct phase fidelity measurements for a single divider.

The board layout is shown in Figure 5.39 with the new pilot tone divider chain magnified in the lower right area. A full block diagram can be found in Section B.3 on page 171. There is only one pilot tone division chain per board to remove the thermal influence of the second chain. Differential noise measurements now have to be performed between two boards. Additional external power splitters, mixers, and filters were installed in the measurement setup that were shown to operate within requirements and are commonly used for these kind of experiments at the *Albert Einstein Institute* in Hanover, Germany.

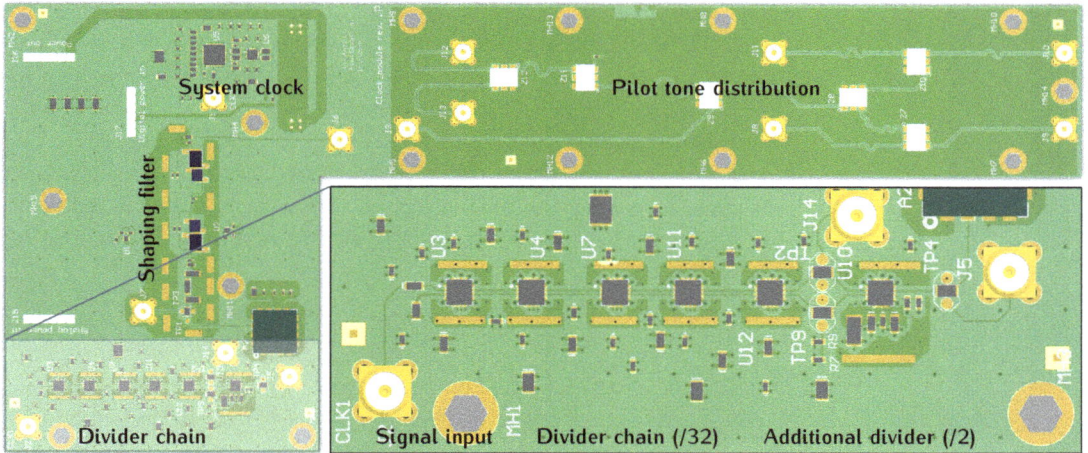

Figure 5.39: Layout of the second revision Frequency Distribution System board. The input signal (J2) is passed to the dividers U5/U6 (top) for clock signal generation and a chain of dividers (U3-4/U7/U10-12) for pilot tone generation. Divider U10 can be bypassed. There is only one pilot tone generation chain per board. Furthermore, the differential GHz phase noise mixer was moved off the board. Voltage stabilization is done separately and all active components can be switched off on demand.

5.4.2.1 PERFORMANCE MEASUREMENTS

All differential measurements presented below were performed between two identical second revision FDS boards. The pilot tone dividers show a significant stability improvement compared to the first revision board (see Figure 5.40, red trace). Timing stability requirements are met over the entire spectrum. Although there is not much room left, the full pilot tone generation and distribution chain still complied with the tight requirements of a timing noise ten times below the carrier read-out noise under certain thermal conditions (blue trace). This measurements included all dividers plus the amplifier, attenuation stage, shaping filter, and the full chain of power splitters for distribution to the ADC boards.

In collaboration with Daniel Edler [116] and with a specially designed breakout board produced by *Axcon Aps* we gained access to the differential system clock signal. The yellow trace in Figure 5.40 shows that it now almost complies even with the strict requirements for the pilot tone, which

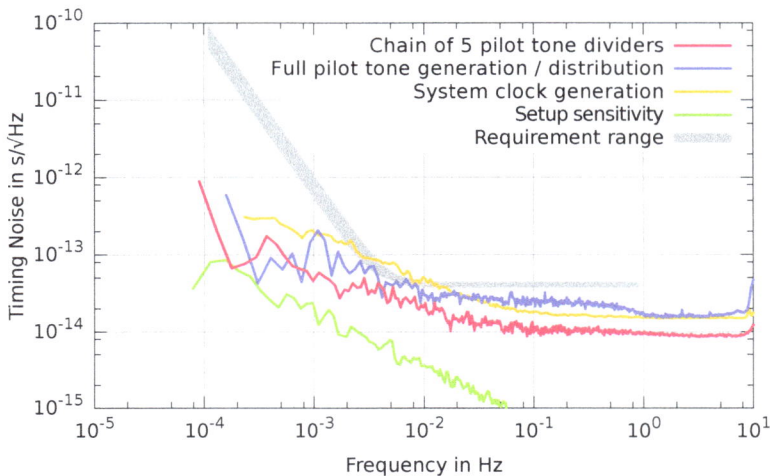

Figure 5.40: Differential phase noise measurements at 75 MHz for the 2nd revision FDS board. The timing stability of pilot tone dividers (red), the full pilot tone generation and distribution chain (blue), and the system clock generation (yellow). The requirement range does not apply to the system clock.

does not apply to the system clock where much higher noise levels could be tolerated. The implementation of the system clock is the same as the pilot tone divider chain of the revision 1 FDS board. Thus the improved performance is compelling evidence that on-board electro-magnetic disturbances and temperature noise by the many heat sources were in fact the limiting factors. Further evidence is the unfortunate fact that the system clock and the pilot tone generation could not be performed on the same FDS board simultaneously. An active system clock divider chain spoiled the phase fidelity of the shaping filter.

*For a full system evaluation that required all signals **simultaneously** we had to use two individual FDS boards, one for the pilot tone, and a separate one for the clock generation.*

5.4.2.2 THERMAL CONSIDERATIONS

For thermal evaluation, the FDS boards were placed inside a custom built actively temperature stabilized housing [132]. Open air temperature fluctuations were below 10^{-1} K/$\sqrt{\text{Hz}}$ at Fourier frequencies of 2×10^{-3} Hz. Additional passive isolation for certain components improved the temperature stability even further down to 10^{-2} K/$\sqrt{\text{Hz}}$. An independent high-precision temperature measurement system based on a Wheatstone bridge with a temperature dependent platinum resistor and a 28-bit ADC front end with subsequent processing performed by an FPGA was used to detect temperature fluctuations near relevant components. Exemplary noise levels are plotted in Figure 5.4 on page 107 (purple and blue traces).

*The **temperature measurement system** is an in-house development by the Albert Einstein Institute*

The temperature stabilized housing allowed us to change the temperature deliberately. For a differential measurement, we placed one FDS board in a thermally stable environment and exposed the other one to a temperature shift from 17 to 21°C. The differential phase of both output signals was measured and we repeated the process multiple times. All electronics relevant for the measurement setup were kept at a constant temperature. Furthermore, we used the most temperature-stable cables available (*Times Microwave Systems* "PhaseTrack210", see Section 5.1.1) to ensure that we only see effects caused by the one board under test. At a frequency of 75 MHz we observed a correlated phase change of 0.5 mrad/K. This corresponds to a coefficient of 1.1×10^{-12} s/K and leads to a temperature stability requirement of at least 3.5×10^{-2} K/$\sqrt{\text{Hz}}$ to meet the timing stability requirement of all considered mission concepts (see table Table 9). This—quite challenging—noise level must be fulfilled throughout the entire FDS board.

*This **phase change** is less severe than the maximum theoretical temperature coefficient of the FDS board's shaping filter (see Section 5.4.1.1) of 1.7×10^{-12} s/K. This means that either the filter is ever so slightly more stable than assumed or that other components on the FDS board feature negative coefficients that counteract the filter's phase shift.*

For our usual differential measurement setup with two FDS boards and all electronics being inside the actively temperature stabilized housing, a shift in temperature still caused a 0.1 mrad/K differential phase change between both output signals. This corresponds to a coefficient of 2.1×10^{-13} s/K and results in a temperature stability requirement of 1.7×10^{-1} K/$\sqrt{\text{Hz}}$ which is still not easy to achieve.

The next step was the integration of the FDS board with the overall metrology system. Since other boards of this system contain many additional ac-

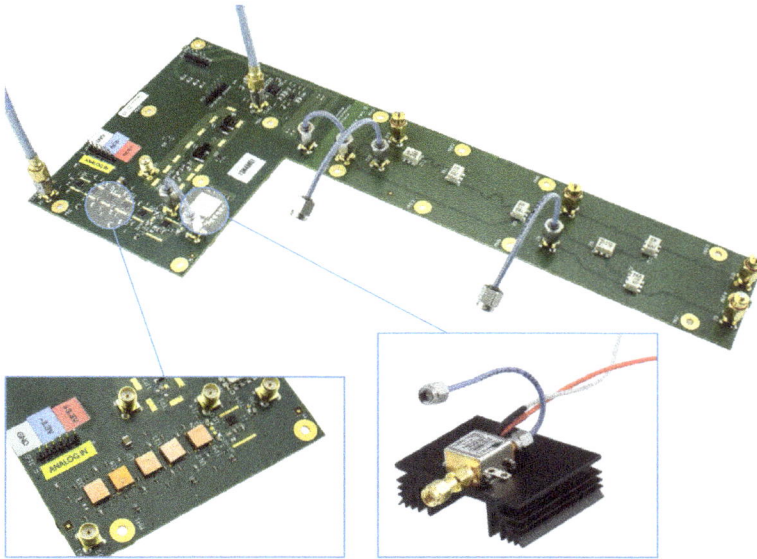

Figure 5.41: FDS board (revision 2) modifications. The original FDS board (top) with incoming 2.4 GHz signals for the pilot tone generation (upper left corner) and system clock division (center top). Unused outputs of the pilot tone distribution are terminated. To reach full performance, the pilot tone dividers were individually shielded by copper caps (bottom left). Furthermore, the on-board amplifier was replaced by a stand alone alternative (bottom right) that was connected to a radiator for better heat dissipation.

tive components, we wanted to further improved the heat dissipation and reduced the influence of electro-magnetic radiation on the FDS board before integration. Two additional modifications are shown in Figure 5.41: We replaced the on-board amplifier by its stand-alone equivalent (*Mini Circuits* "ZX60-33LN-S"). It was equipped with SMA connectors, shielded by a rigid housing, and connected to a radiator to dissipate heat. Custom-milled copper filter caps with a wall thickness of 0.5 mm where placed on top of the first five dividers of the pilot tone generation chain. Thermally conductive foam connected the divider surface with the inside of the cap so that head would not be trapped within the cap. Additionally, we observed a phase shift when humans (or other large capacities) moved nearby the shaping filter. Thus we connected it thermally to copper foil (shielding the filter) and a radiator (to ensure a good thermal coupling to the temperature stable environment).

5.4.3 INTEGRATION

The final Frequency Distribution System board was integrated with the modular phasemeter that was designed and tested by the international "LISA Metrology System Team" as part of a technology development activity by the *European Space Agency*. A single DAC card was built that is responsible for laser locking and offset frequency switching. The differential system clock with 50% duty cycle was produced by the FDS board from a 2.4 GHz sine signal and passed to the main processing module via a proprietary connector on the bottom of the board. The pilot tone was generated from the same signal and connected to five ADC cards by cable bridges. Each ADC card can handle up to four input channels with multiple signals (pilot tone, carrier, and sideband beat notes) per channel. The pilot tone is added to each channel on the corresponding ADC module. Figure 5.42 shows the fully assembled metrology system.

Figure 5.42: Full LISA Metrology System with main board (below), DAC module (bottom left) and Frequency Distribution System board (center) connected to 5 ADC Modules.

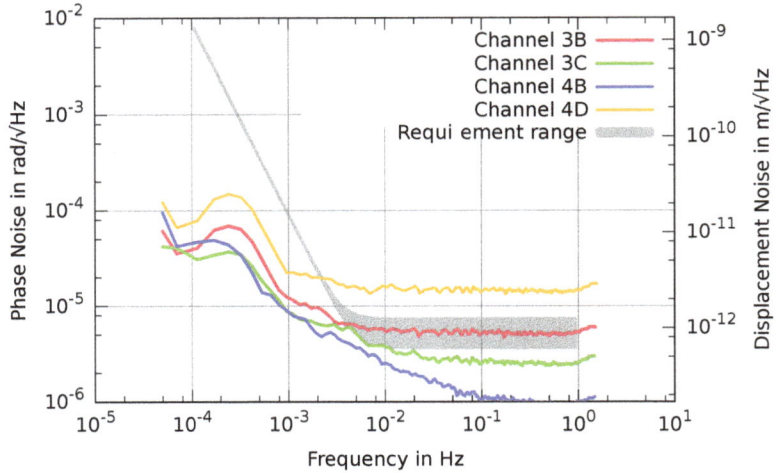

Figure 5.43: Phase noise between two ADC cards measured by the phasemeter that was developed in parallel with the FDS board. The system clock was provided by the FDS board. Noise in the different channels depends on impedance matching. A good matching complies with all requirements. A rather poor matching can introduce a phase noise may be incompatible with the proposed missions Channel 4D was purposefully unmatched [132].

This **realistic signal** included carrier and sideband beat notes at an SNR of 75 dB Hz (inter-spacecraft interferometer), DS/SS modulation (PRN code), simulated Dopper shifts, and correct levels of laser frequency and amplitude noise.

For performance measurements of the phasemeter the FDS board provided the differential system clock and the pilot tone. An FPGA based digital signal simulator [74] generated a realistic signal that was split and passed to different ADC channels. Adders on the ADC cards injected the pilot tone to each channel. We determined the differential phase noise between 25 MHz carrier beat notes of two channels on different ADC cards and used the pilot tone for ADC jitter correction (see Section 4.3.1). Results are plotted in Figure 5.43. For sufficiently good impedance matching between the input sig-

Figure 5.44: Full pilot tone generation and transmission chain. The Frequency Distribution System board (green), cable bridges (orange), and GHz electronics with power splitters and mixers (purple) are necessary to reach performance between different ADC cards. For interspacecraft synchronization, GHz modulation and PRN codes need to be imprinted onto the outgoing laser beams by EOMs (blue) and amplified by fiber amplifiers (red). Timing noise for all components in the pilot tone generation and transmission chain is shown in Figure 5.45. Temperature coefficients for signal lines and relevant components can be found in Table 13.

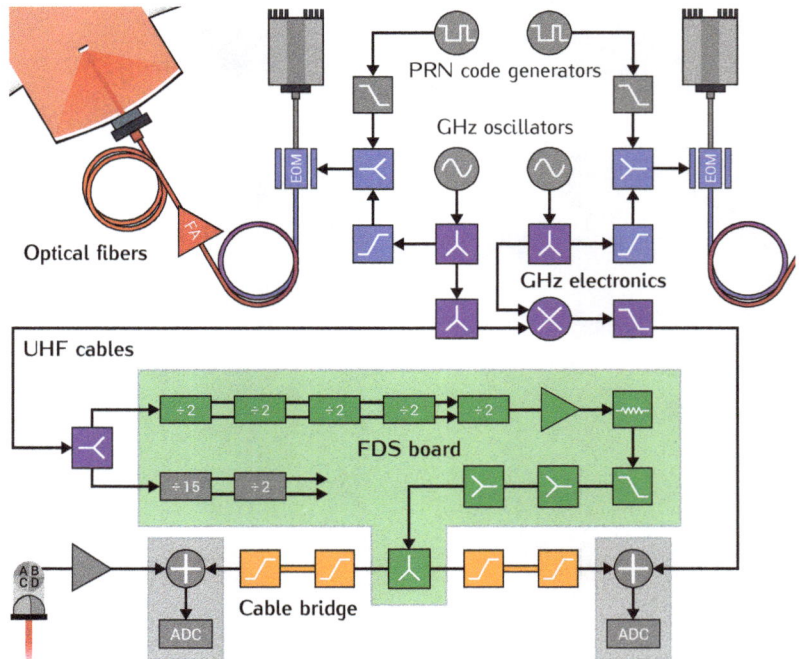

nals even the toughest phase noise requirements (see Table 9) are met [132]. This implies that

✦ the ADC jitter correction scheme works since otherwise intrinsic timing noise between the two different ADCs would dominate,

✦ the pilot tone adders meet the timing stability requirements, and

✦ all signals provided by the FDS board are compatible with the phasemeter and the overall metrology system.

This test only shows the phase fidelity of the pilot tone distribution part (power splitters, signal lines, and cable bridges) of the FDS board. When you want to achieve the same level of performance between different phasemeters that have independent system clocks (or are located on different spacecraft), the full pilot tone generation and transmission chain is required. The currently implemented design is illustrated in Figure 5.44. Individual components are grouped in colors. Each component except for gray ones has to meet the timing stability requirements so that differential clock jitter can be successfully removed by an adequate post processing technique. Based on the measurements presented in this chapter we can predict that this is in fact the case.

Timing jitter for all component are summarized in Figure 5.45. Colors match the component groups in Figure 5.44. The green trace combines the performance of the entire Frequency Distribution System board (pilot tone

Figure 5.45: Timing jitter of the entire pilot tone generation and transmission chain, shown for individual component groups. Colors match the groups in Figure 5.44. All components comply with the timing stability requirements of ten times below the carrier read-out noise level for all considered mission concepts.

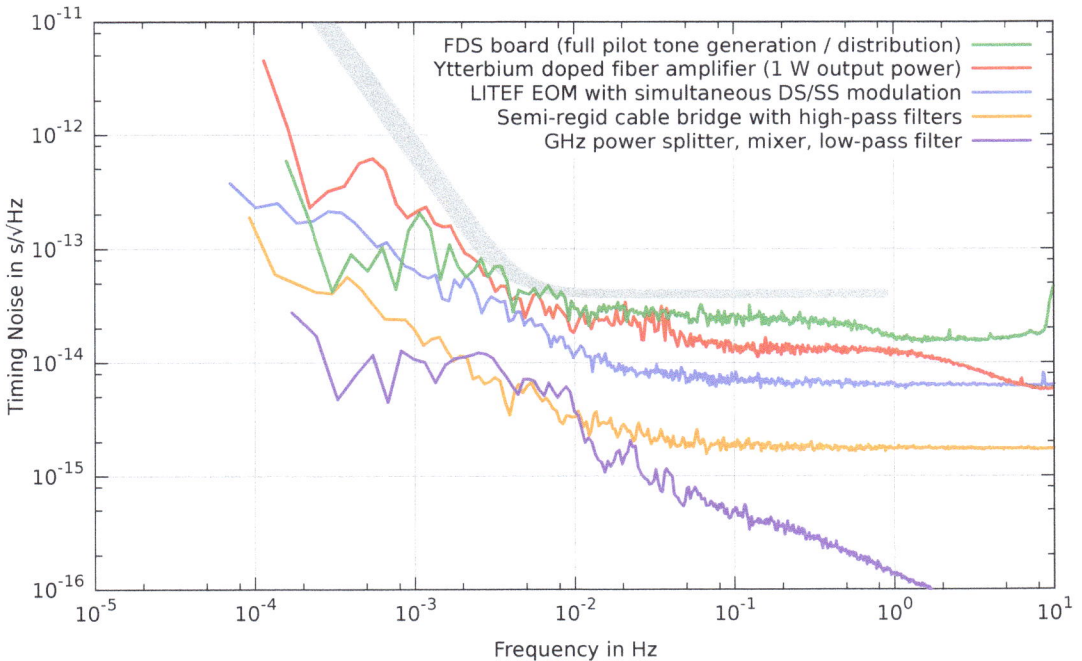

chain, see Section 5.4.2.1). The red one shows the performance for an exemplary laser fiber amplifier at 1 W output power in combination with optical fibers that were necessary for the test setup (see Section 5.1.4). The blue trace is the performance of the *LITEF* EOM with attached fibers and in combination with noise induced by the UHF cable carrying the modulation frequency (see Section 5.1.2.1). The orange one combines the semi-regid cable bridges that were already part of the differential phase noise measurement (see Figure 5.43) with the high-pass filters to suppress crosstalk (see Section 5.3.3). The purple trace finally comprises all GHz electronics such as power splitters, mixers, and filters (see Section 5.3.4). All components comply with the timing stability requirements. The combined timing jitter for the entire pilot tone generation and transmission chain may exceed the required level. However, one should keep in mind that this level is a factor of ten below the carrier read-out noise. This ensures that even the combination of different noise sources has no significant impact on the observatory's detection limit.

The timing noise of some of the components mentioned above highly depends on the exact type, absolute temperature, and overall temperature stability. The worst-case temperature coefficients for critical components are shown in Table 13.

Table 13: Temperature coefficient for different components.

*depends on jacketing
**depends on temperature
***depends on cable type and temperature

Component	Coefficient $\Delta t/\Delta T$	Range	For details see
Optical fibers (1 m)*	$\sim 5 \times 10^{-14}$ s/K	22...29°C	Section 5.1.3
EOM**	$< 4.5 \times 10^{-13}$ s/K	16...29°C	Section 5.1.2.2
UHF cables (1 m)***	$< 1.2 \times 10^{-12}$ s/K	5...50°C	Section 5.1.1
FDS board	1.1×10^{-12} s/K	17...21°C	Section 5.4.2.2

Detailed knowledge of the temperature stability at different parts of the spacecraft is necessary to ensure that the timing stability requirements can be met. The same is true for laboratory setups. This includes temperature noise generated by active components of the metrology system such as ADCs and FPGAs. Critical components should be places on individual boards separated from active parts if possible.

In conclusion, **we made a viable constellation-wide reference signal scheme available for the very first time**. Signals generated by the Frequency Distribution System board comply with the required signal shape, power levels, and pilot tone phase fidelity. All additional components of the megahertz signal path that are not located on the FDS board were tested successfully in Section 5.3. Additionally, in Section 5.1 we could show that UHF cables, EOMs, fiber amplifiers, and optical fibers exist that comply with all requirements of the full pilot tone generation and transmission chain.

Some components were only tested differentially. This means that some noise sources might not have shown up due to common mode noise suppression. Only a '3-signal-test' that involves three different frequencies would reveal the full extend of correlated noise sources and nonlinearities. The most reliable way to measure the performance of the Frequency Distribution System board would of course be in a realistic setup that simulates the actual clock tone transfer and clock noise correction with optical signals between different and independent metrology systems. The presented results lay the groundwork for such tests that will be performed in the upcoming years at the *Albert Einstein Institute* in Hanover, Germany. The already mentioned hexagonal electro-optical test stand [133] required for such test is currently under construction. ∎

Part IV

CONCLUSION

Space missions take a long time to be developed and tested. This is particularly true for a spaceborne gravitational wave observatory which requires demanding technologies. Most of these technologies were already developed on ground. Others, such as the gravity reference sensors, will soon be tested during the 'LISA pathfinder' mission. This leaves the metrology system including the Frequency Distribution System and related technologies as the main milestone which needs to be completed before the first LISA mission is ready to launch.

The Frequency Distribution System was designed, developed, and tested as part of this thesis. It complies with the timing stability requirements for all considered mission concepts. These requirements were re-evaluated in detail including a thorough study of the maximum heterodyne frequencies considering the technical limitations by current laser systems and different locking schemes in combination with a sophisticated offset frequency switching plan.

Now, all systems have to be qualified for space applications and evaluated with optical signals in a realistic environment. Enough time remains to finalize the technology for a targeted launch of a interferometric gravitational wave observatory in 2034. This first observatory will herald the beginning of a new and exciting era in space exploration and will be a revolutionary step forward in astronomy, cosmology, and fundamental physics alike.

6

The developed 'Gravitational Wave Observatory Designer' provided the toolset to re-evaluate different system requirements. An important piece of the puzzle was the as yet unknown heterodyne frequency range of the many different mission concepts. For the first time, minimum and maximum beat note frequencies were determined for various arm lengths, laser relative intensity noise levels, and mission durations. Current laser systems do not allow heterodyne frequencies below 5 MHz. Individual mission concepts require maximum heterodyne frequencies between 10 and 28 MHz. Taking this into account, I could show that designs exist which impose less tough restrictions on phase fidelity and timing stability, yet result in observatories that yield an equally excellent sensitivity. However, more detailed mission studies are required to fully evaluate such concepts. Additionally, orbits could be further optimized with algorithms that directly include the presented locking schemes.

Throughout this thesis I made use of the more demanding set of requirements that result from currently considered missions. All concepts have in common that available reference oscillators and analog-to-digital (ADC) converters can not directly meet the required timing stability. Without a solution to this issue, the combination of independent measurements from different spacecraft would introduce an equivalent displacement noise orders of magnitudes above the design sensitivity.

The only verified solution to this problem is an Inter-Spacecraft Frequency Distribution System as described in this thesis. It uses a GHz modulation signal that is imprinted onto the outgoing laser beams to be transmitted to the distant spacecraft. This enables a constellation wide synchronization of measurements. Locally, the GHz modulation is converted to a phase stable MHz signal—the 'pilot tone'—that is superimposed onto the local beat note signals to suppress additional ADC timing jitter. The two auxiliary signals – sideband beat notes and the pilot tone – which are present in every ADC channel suffice to suppress any timing jitter during data processing below the required timing stability. Of course, this scheme only works for a sufficient phase fidelity between different representations of the reference frequency. Thus the timing stability of each component in the entire signal line needs to be sufficiently low.

The developed Inter-Spacecraft Frequency Distribution System is fully functional. Starting with a GHz oscillator, we demonstrated that components exist that meet all functional and performance requirements to build a complete pilot tone generation, conversion, distribution, and transmission chain. To transmit the GHz signal between spacecraft, we found that

+ standard UHF cables may not be a good choice for gravitational wave observatories since they show a large temperature coefficient at room temperatures, but alternatives exist that meet all requirements,

+ a radiation hard and vacuum compatible electro-optic modulator by *LITEF* easily fulfills the phase fidelity requirements, offers a damage threshold of up to 50 mW, and features a high efficiency and moderate temperature coefficient,

+ fiber amplifiers exist that comply with the phase fidelity requirements for output powers of up to 1 W, and do not add excess RIN but are instead limited by the shot noise level of the seed laser, and that

+ optical fibers generally have a very low temperature coefficient and thus can carry the modulated laser light to the optical bench with sufficient phase fidelity.

In the other direction, from the GHz modulation signal towards the ADC front end, we

+ identified digital dividers that 1) can be used in sequence to divide the GHz modulation signal down to a MHz signal, 2) do not violate the required timing stability, and 3) are compatible with the overall metrology system,

+ found that electrical amplifiers and attenuation stages with sufficient timing stability exist,

+ could design a custom shaping filter with an excellent temperature coefficient to suppresses higher order modes in the MHz signal,

+ identified power splitters that distribute the pilot tone to the different ADC cards without adding limiting excess phase noise.

On top of that, all systems for auxiliary functions – like mixers for the differential GHz oscillators signal or power combiners for the modulation signal and the PRN code – were designed and tested successfully. A differential system clock can be generated out of the same GHz reference as well. Our Inter-Spacecraft Frequency Distribution System completes the metrology system development for spaceborne gravitational wave observatories. This system was for a long time the last missing important technology item in Europe relevant for gravitational wave observatories. It is now finally available, complies to the Technology Readiness Level 4 specifications, and represents a huge milestone on the way towards the 2034 launch date.

However, the work on the Inter-Spacecraft Frequency Distribution System is not done yet. There are only a few more years before the actual construction of the spacecraft will begin, and there are many more milestones ahead. The already introduced '3-signal-test' will reveal any correlated noise sources and nonlinearities. The final goal is a representative metrology test bed: different independent metrology systems that act on realistic optical signals will test every single component in the entire reference signal path to full extend. With the same setup, noise suppression algorithms can be evaluated in detail with realistic data streams. Additionally, an optical test bed would open the possibility for an optical pilot tone. This concept has never been demonstrated before. Here, the reference signal is not augmented electronically to the heterodyne signal directly in front of the ADCs, but optically superimposed onto the light received by the photodetector. Such a pilot tone cannot only cancel system clock and ADC timing jitter, but also removes excess phase noise caused by the photoreceiver electronics. The potential benefits are very tempting and results of these tests will be highly anticipated.

Meanwhile, the metrology system needs to be developed further in an iterative process. At the moment, it produces an abundance of heat that might in the end violate the temperature stability requirements of the FDS board. Further thermal engineering will be required to design a custom housing that allows for a stable operation of the entire system. Less powerful FPGAs and in general more efficient electronics that go hand in hand with space qualified hardware are necessary. This will not only help to reduce issues with heat dissipation but also lower the overall power consumption of the metrology system.

As time progresses, future revisions of the frequency conversion and transmission chain will aim for higher TRL levels. So far, only a few components were actually space qualified. The *LITEF* EOM and some UHF cables we tested are the exceptions here. Other components might be easy to qualify for space applications. Yet, all radiation hard dividers under test failed the timing stability requirements. Another market research that concentrates on the additional requirements demanded of space missions will be needed to identify components that can replace currently used electronics. In the end, the whole system has to be ready for the actual space mission.

Considerable work will be necessary to produce real flight hardware. But once this is done, humanity will be ready for the next giant leap in science. Soon, the first gravitational wave observatory will go into space and start unraveling the mysteries of the universe. We will not only prove that gravitational waves exist as Einstein predicted. We will be doing detailed measurements of astrophysical phenomena that have never been possible before by any other means. In this spirit, let me close with a quote by Galileo Galilei...

Figure 6.1: Rendering of one spacecraft of the Classic LISA gravitational wave observatory mission concept.
credit: Vladimir Arndt/iStock (galaxy), Kevin Carden/123RF (background), AEI/MM/Exozet (spacecraft)

"Measure what is measurable,
and make measurable what is not so."
— Galileo Galilei

Part V

APPENDIX

SOURCE CODE

A.1 GENETIC ALGORITHM

Source code and inter-spacecraft Doppler frequencies in the line of sight available at http://github.com/gulbrillo/genetic-frequencies.

A.1.1 MAIN LOOP

```perl
#!/usr/bin/perl -w

use strict;
use List::Util qw[min max];
use Switch;
$|++; #unbuffer stout

#=== CONFIG START

#MISSION PARAMETERS
our $duration = 5; #mission duration $duration = 2, 5, 10 years
our $arm = 5; #armlength $arm = 1, 2, 3, 5, 5 million kilometers
#requires file name '"doppler_".$arm.".0.dat"' in folder '"/"$duration."year/"' with
    Doppler frequencies fd1, fd2, fd3 (one column per day) in Hz
our $links = 6; #number of $links = 4, 6

our $min = 7e6; #minimum frequency in Hz
our $max = 20e6; #maximum frequency in Hz

our $mode = 1; #1: no permutation of master laser. 2: do one run for each master laser
    position (6 runs). 3: find optimum maser laser position for each segment.
our $permutation = 123; #123 to 321 (only for $mode=1)
our $scheme = 1; #locking scheme $scheme = 1 (A), 2 (B), 3 (C)

#GENETIC PARAMETERS
our $population = 200; #population size
our $generations = 20; #number of generations
our $percentage = 50; #number of individuals allowed to procreate in percent
our $mutatingprob = 10; #mutation probability in percent
our $mutatingbits = 16; #least significant bits to mutate
our $recombination = 60; #probability that two offset frequencies are combined percent
    instead of one of both being used unchanged
our $alive = 1; #all offspring needs to survive for at least 1 day. 0: yes. 1: no.

#DEBUG
our $printdebug = 1; #write iterations for each generation. 0: no. 1: yes.
our $debugfolder = "debug"; #folder for debug files

#OUTPUT FILE DIRECTORY
my $mydate = &datum_zeit();
our $filedir = $mydate."_mission$duration-arm$arm"."_pop$population-gen$generations-
    per$percentage-int$recombination-mut$mutatingprob.$mutatingbits-min$min-max$max-
    scheme$scheme/";
```

```perl
#=== CONFIG END

my $exec;
$exec = `mkdir $filedir `;
$debugfolder = $filedir."debug";
$exec = `mkdir $debugfolder `;

our $filename = "";
our $switchmaster = 0;
our $run = 1;

switch ($mode) {
    case 1 {$switchmaster = 0; $run = 1;}
    case 2 {$switchmaster = 0; $run = 6;}
    case 3 {$switchmaster = 1; $run = 1;}
}

our $permutations;
if ($switchmaster == 1) {$permutations = 6;} else {$permutations = 1;}

our $A;
our $B;
our $C;
our @BN111; #beat-note between lasers of S/C1 on S/C1
our @BN121; #beat-note between lasers of S/C1 and S/C2 on S/C1
our @BN122; #beat-note between lasers of S/C1 and S/C2 on S/C2
our @BN131;
our @BN133;
our @BN222;
our @BN333;
our @BN232;
our @BN233;

our $end = 0;

#LOAD FILE
my $dopplerfile = $duration."year/doppler_".$arm.".0.dat";
my @FILE;
my $i = 0;
our @D; #Doppler shifts: ${$D[1]}[$day], ${$D[2]}[$day], ${$D[3]}[$day]

open (FILE, "$dopplerfile") or die "Cannot open $dopplerfile for reading: $!\n";
@FILE = <FILE>;
close FILE;

my $row;

foreach $row (@FILE) {

    if ($row !~ /^#/) {
        $end++;
        my @ROW;
        $row =~ s/\n//g;
        @ROW = split("\ +|\t+",$row);
        my $value;
        $i = 0;
        foreach $value (@ROW) {
            $i++;
            push(@{$D[$i]}, $value);
        }
    }

}
```

```perl
     for (my $thisrun = 0; $thisrun < $run; $thisrun++) {

106      if ($run > 1) {
             switch ($thisrun) {
                 case 0 {$A = 1; $B = 2; $C = 3;}
                 case 1 {$A = 3; $B = 2; $C = 1;}
                 case 2 {$A = 2; $B = 3; $C = 1;}
111              case 3 {$A = 1; $B = 3; $C = 2;}
                 case 4 {$A = 3; $B = 1; $C = 2;}
                 case 5 {$A = 2; $B = 1; $C = 3;}
             }
         }
116
         #output file name:
         switch ($mode) {
             case 1 {$filename = $filedir."$permutation";}
             case 2 {$filename = $filedir."$A$B$C";}
121          case 3 {$filename = $filedir."optimum";}
         }

         #initiate output files
         my $OutF = $filename.".txt";
126      open(OUTPUT, ">$OutF") or die "Can't open or create file $OutF: $!\n";
         print OUTPUT "#DAY #BN11\@1 #BN12\@2 #BN12\@1 #BN13\@3 #BN13\@1 #BN22\@2 #BN33\@3 #
             BN23\@2 #BN23\@3\n";
         close(OUTPUT);

         my $BrkF = $filename.".brk";
131      open(BREAKS, ">$BrkF") or die "Can't open or create file $OutF: $!\n";
         print BREAKS "#Breaks\n";
         close(BREAKS);

         my $FrqF = $filename.".frq";
136      open(FRQ, ">$FrqF") or die "Can't open or create file $FrqF: $!\n";
         print FRQ "#Frequency plan\n";
         close(FRQ);

         my $LOCK13 = 0; #lock between master laser and 2nd laser on S/C1
141      my $LOCK21 = 0;
         my $LOCK31 = 0;
         my $LOCK23 = 0;
         my $LOCK32 = 0;

146      my $start = 1; #start at day 1
         my $from = -$max;
         my $to = $max;

         print "Simulation for $end days.\n\n";
151
         open(FRQ, ">>$FrqF") or die "Can't open or create file $FrqF: $!\n";
         print FRQ "#Simulation for $end days.\n# min = $min;\n# max = $max;\n# from = $from
             ;\n# to = $to;\n";
         print FRQ "#day LOCK13 LOCK21 LOCK31 LOCK23 LOCK32 days nice\n";
         close(FRQ);
156
         while ($start < $end) #try to keep beat-notes within frequency range for as long as
             possible. stop when end of file.
         {
             my @F = ();
             my @Fmax = ();
161          my $dimensions = 5;
             my @POP;
             my $maxnice = 0;
```

```perl
        for (my $p = 0; $p<$permutations; $p++) {
            if ($run==1) {
                if ($switchmaster == 1) {
                    switch ($p) {
                        case 0 {$A = 1; $B = 2; $C = 3;}
                        case 1 {$A = 3; $B = 2; $C = 1;}
                        case 2 {$A = 2; $B = 3; $C = 1;}
                        case 3 {$A = 1; $B = 3; $C = 2;}
                        case 4 {$A = 3; $B = 1; $C = 2;}
                        case 5 {$A = 2; $B = 1; $C = 3;}
                    }
                } else {
                    switch ($permutation) {
                        case 123 {$A = 1; $B = 2; $C = 3;}
                        case 321 {$A = 3; $B = 2; $C = 1;}
                        case 231 {$A = 2; $B = 3; $C = 1;}
                        case 132 {$A = 1; $B = 3; $C = 2;}
                        case 312 {$A = 3; $B = 1; $C = 2;}
                        case 213 {$A = 2; $B = 1; $C = 3;}
                    }
                }
            }

            @POP = ();

            #GENERATE GENERATION 0
            for (my $p = 0;$p<$population;$p++) {
                my $last = 0;
                my $nice = 0;
                my $warning = 0;
                my @sign = (-1,1); #positive or negative
                while ($last-$start < $alive) {
                    $F[0] = $sign[int(rand 2)]*(rand($max-$min)+$min)/1e6;
                        $F[1] = $sign[int(rand 2)]*(rand($max-$min)+
                            $min)/1e6;
                        $F[2] = $sign[int(rand 2)]*(rand($max-$min)+
                            $min)/1e6;
                        $F[3] = $sign[int(rand 2)]*(rand($max-$min)+
                            $min)/1e6;
                        $F[4] = $sign[int(rand 2)]*(rand($max-$min)+
                            $min)/1e6;
                    ($last,$nice) = beatnotes($start, $F[0], $F[1], $F[2], $F[3], $F
                        [4]);
                    $warning++;
                    if (int($warning/100000) == $warning/100000) {print "WARNING:
                        CANNOT FIND ANY COMBINATION IN $warning TRYS!\n"}
                }

                #nice in row 7: $last-$start-$nice. $last-$start is number of stable
                    days, $nice is the mean value of the distance of the beat-notes
                    from the center
                push (@POP, [$F[0], $F[1], $F[2], $F[3], $F[4], $last, $nice, $last-
                    $start-$nice]);

                #debug:
                if ($printdebug) {
                    my $debugfile = "$debugfolder/day$start-$A$B$C.go";
                    open(DEBUG, ">>$debugfile") or die "Can't open or create file
                        $debugfile: $!\n";
                    print DEBUG $start." ".$F[0]." ".$F[1]." ".$F[2]." ".$F[3]." ".$F
                        [4]." ".($last-$start)."\n";
                close(DEBUG); }
            }
```

```perl
@POP = sort { $a->[7] <=> $b->[7] } @POP; #sort by $last-$start-$nice
print "Day $start \@$A$B$C: stable for ".($POP[$population -1][5]-$start)."
    days, nice: $POP[$population -1][6] ($POP[$population -1][0], $POP[
    $population -1][1], $POP[$population -1][2], $POP[$population -1][3], $POP
    [$population -1][4])\n";

#EVOLUTION
my $selection = int($population*$percentage/100); #take the best of the
    best

for (my $e = 0;$e<$generations;$e++) { #number of generations

    my @POPnew = ();

    #produce the next generation
    for (my $p = 0;$p<$population -1;$p++) {

        my $last = 0;
        my $nice = 0;
        my @Fnew = ();

        my $male = (int(rand($selection))+$population-$selection); #choose
            a male partner

        while ($last-$start < $alive) {

            my $female = (int(rand($selection))+$population-$selection); #
                choose a female partner

            my $negativ;
            my $cut;
            my $cut1;
            my $cut2;
            my $femalev;
            my $malev;
            my $interaction;
            my $mutation;
            my $mut;

            #PROCREATE
            for (my $g = 0;$g<5;$g++) { #for all 5 offset lock frequencies

                $interaction = int(rand(100));
                if ($interaction < $recombination) { #Interaktionsrate in %

                    #use most significant bits of the father
                    if ($POP[$male][$g] < 0) {$negativ = -1} else {$negativ
                        = 1}
                    $femalev = abs($POP[$female][$g]) * 100000 +0.5;
                    $malev = abs($POP[$male][$g]) * 100000 +0.5;

                    #where to cut the gene
                    $cut = int(rand 17)+1;
                    $cut1 = 2**($cut)-1;
                    $cut2 = 0b111111111111111111111111111111 << $cut;

                    #cut male and female gene
                    $femalev = $femalev & $cut1;
                    $malev = $malev & $cut2;

                    #offspring
                    $Fnew[$g] = $femalev | $malev;
                    $Fnew[$g] = $Fnew[$g] / 100000 * $negativ;
```

```perl
                                    } else {
                                        $Fnew[$g] = $POP[$male
                                            ][$g];
                                    }

                    $mutation = int(rand(100));
                    if ($mutation < $mutatingprob) { #mutation rate in %
                        if ($Fnew[$g] < 0) {$negativ = -1} else {$negativ = 1}
                        $mut = 1 << int(rand $mutatingbits); #mutation: flip a
                            bit somewhere
                        $Fnew[$g] = abs($Fnew[$g]) * 100000;
                        $Fnew[$g] = $Fnew[$g] ^ $mut;
                        $Fnew[$g] = $Fnew[$g] / 100000 * $negativ;
                    }
                }
                ($last, $nice) = beatnotes($start, $Fnew[0], $Fnew[1], $Fnew[2],
                    $Fnew[3], $Fnew[4]);
            }

            push (@POPnew, [$Fnew[0], $Fnew[1], $Fnew[2], $Fnew[3], $Fnew[4],
                $last, $nice, $last-$start-$nice]);

            #debug:
            if ($printdebug) {
                my $debugfile = "$debugfolder/day$start-$A$B$C.g".($e+1);
                open(DEBUG, ">>$debugfile") or die "Can't open or create file
                    $debugfile: $!\n";
                print DEBUG $start." ".$Fnew[0]." ".$Fnew[1]." ".$Fnew[2]." ".
                    $Fnew[3]." ".$Fnew[4]." ".($last-$start)."\n";
                            close(DEBUG);
                }
        }
        #make sure that the strongest individual from the last generation is
            part of the new generation
        push (@POPnew, [$POP[$population-1][0], $POP[$population-1][1], $POP[
            $population-1][2], $POP[$population-1][3], $POP[$population-1][4],
            $POP[$population-1][5], $POP[$population-1][6], $POP[$population
            -1][7]]);

        if ($printdebug) {
            my $debugfile = "$debugfolder/day$start-$A$B$C.g".($e+1);
            open(DEBUG, ">>$debugfile") or die "Can't open or create file
                $debugfile: $!\n";
            print DEBUG $start." ".$POP[$population-1][0] ." ".$POP[$population
                -1][1] ." ".$POP[$population-1][2] ." ".$POP[$population-1][3]
                ." ".$POP[$population-1][4] ." ".($POP[$population-1][5]-$start
                )."\n";
                            close(DEBUG);
                }

        @POP = @POPnew; #replace generation

        @POP = sort { $a->[7] <=> $b->[7] } @POP; #sort by $last-$start-$nice
        print "Generation ".($e+1).": stable for ".($POP[$population-1][5]-
            $start)." days, nice: $POP[$population-1][6] ($POP[$population
            -1][0], $POP[$population-1][1], $POP[$population-1][2], $POP[
            $population-1][3], $POP[$population-1][4])\n";

    }

    if ($POP[$population-1][7] > $maxnice) {
        $maxnice = $POP[$population-1][7];
        $permutation = $A.$B.$C;
        $Fmax[0] = $POP[$population-1][0];
```

```
                    $Fmax[1]  =  $POP[$population -1][1];
                    $Fmax[2]  =  $POP[$population -1][2];
                    $Fmax[3]  =  $POP[$population -1][3];
                    $Fmax[4]  =  $POP[$population -1][4];
326                 $Fmax[5]  =  $POP[$population -1][5];
                    $Fmax[6]  =  $POP[$population -1][6];
                }

            }
331
            switch ($permutation) {
                case 123 {$A = 1; $B = 2; $C = 3;}
                case 321 {$A = 3; $B = 2; $C = 1;}
336             case 231 {$A = 2; $B = 3; $C = 1;}
                case 132 {$A = 1; $B = 3; $C = 2;}
                case 312 {$A = 3; $B = 1; $C = 2;}
                case 213 {$A = 2; $B = 1; $C = 3;}
            }
341
            #write the optimum frequency combination to file
            open(FRQ, ">>$FrqF") or die "Can't open or create file $FrqF: $!\n";
            print FRQ $start." ".$Fmax[0]." ".$Fmax[1]." ".$Fmax[2]." ".$Fmax[3]." ".$Fmax
                [4]." ".($Fmax[5]-$start)." ".$Fmax[6] ." ".$A.$B.$C."\n";
346         close(FRQ);

            print "> ".($Fmax[5]-$start)." days \@$permutation!\n\n";

            #calculate the beat-notes
351         &beatnotes($start, $Fmax[0], $Fmax[1], $Fmax[2], $Fmax[3], $Fmax[4]);

            #write the beat-notes to file
            &ausgabe($start,$Fmax[5]);

356         $start = $Fmax[5]; #new starting point (day)
        }

    }

361
    $|++; #unbuffer (flush) STOUT

    exit 0;
```

A.1.2 CALCULATE BEAT-NOTE FREQUENCIES

```
1   sub beatnotes {

    my $start = shift;
    my $LOCK13 = shift;
    $LOCK13 = $LOCK13 * 1e6;
6   my $LOCK21 = shift;
    $LOCK21 = $LOCK21 * 1e6;
    my $LOCK31 = shift;
    $LOCK31 = $LOCK31 * 1e6;
    my $LOCK23 = shift;
11  $LOCK23 = $LOCK23 * 1e6;
    my $LOCK32 = shift;
    $LOCK32 = $LOCK32 * 1e6;
```

```perl
     my $i; #day to start at
16   my $nice = 0;
     my $n = 0;
     for ($i=$start;$i<$end;$i++)
     {
         switch ($scheme) {
21           case 1 {
                 #LOCKING SCHEME A
                 $BN111[$i] = $LOCK13;
                 $BN122[$i] = $LOCK21;
                 $BN121[$i] = 2*${$D[$A]}[$i]+$LOCK21;
26               $BN133[$i] = $LOCK31;
                 $BN131[$i] = 2*${$D[$C]}[$i]+$LOCK31;
                 $BN222[$i] = $LOCK23;
                 $BN333[$i] = $LOCK32;
                 $BN232[$i] = ${$D[$A]}[$i]+$LOCK21+$LOCK23-($LOCK13+${$D[$C]}[$i]+
                     $LOCK31+$LOCK32+${$D[$B]}[$i]);
31               $BN233[$i] = ${$D[$A]}[$i]+$LOCK21+$LOCK23+${$D[$B]}[$i]-($LOCK13+${$D[
                     $C]}[$i]+$LOCK31+$LOCK32);

                 if ($links < 6) {
                                     $BN222[$i] = $min+0.5;
                                     $BN333[$i] = $min+0.5;
36                                   $BN232[$i] = $min+0.5;
                                     $BN233[$i] = $min+0.5;
                                 }
             }
         case 2 {
41               #LOCKING SCHEME B
                 $BN111[$i] = ${$D[$A]}[$i]+${$D[$B]}[$i]+${$D[$C]}[$i]+$LOCK21+$LOCK23+
                     $LOCK32+$LOCK31+$LOCK13;
                 $BN122[$i] = $LOCK21;
                 $BN121[$i] = 2*${$D[$A]}[$i]+$LOCK21;
                 $BN133[$i] = 2*${$D[$C]}[$i]+$LOCK13;
46               $BN131[$i] = $LOCK13;
                 $BN222[$i] = $LOCK23;
                 $BN333[$i] = $LOCK31;
                 $BN232[$i] = 2*${$D[$B]}[$i]+$LOCK32;
                 $BN233[$i] = $LOCK32;

51
             }
         case 3 {
                             #LOCKING SCHEME C
                 $BN111[$i] = $LOCK13;
56               $BN121[$i] = 2*${$D[$A]}[$i]+$LOCK21;
                 $BN131[$i] = ${$D[$A]}[$i]+${$D[$B]}[$i]+${$D[$C]}[$i]+$LOCK21+$LOCK23+
                     $LOCK32+$LOCK31-$LOCK13;
                 $BN222[$i] = $LOCK23;
                 $BN122[$i] = $LOCK21;
                 $BN232[$i] = 2*${$D[$B]}[$i]+$LOCK32;
61               $BN333[$i] = $LOCK31;
                 $BN133[$i] = $LOCK13+${$D[$C]}[$i]-${$D[$A]}[$i]-$LOCK21-$LOCK23-${$D[
                     $B]}[$i]-$LOCK32-$LOCK31;
                 $BN233[$i] = $LOCK32;
             }
         }
66
         $nice = $nice + (abs(abs($BN111[$i])-(($min+$max)/2))*abs(abs($BN111[$i])-((
             $min+$max)/2)) + abs(abs($BN122[$i])-(($min+$max)/2))*abs(abs($BN122[$i])
             -(($min+$max)/2)) + abs(abs($BN121[$i])-(($min+$max)/2))*abs(abs($BN121[$i
             ])-(($min+$max)/2)) + abs(abs($BN133[$i])-(($min+$max)/2))*abs(abs($BN133[
             $i])-(($min+$max)/2)) + abs(abs($BN131[$i])-(($min+$max)/2))*abs(abs($BN131
             [$i])-(($min+$max)/2)) + abs(abs($BN222[$i])-(($min+$max)/2))*abs(abs(
```

```
        $BN222[$i]) −(($min+$max)/2)) + abs(abs($BN333[$i]) −(($min+$max)/2))*abs(abs
        ($BN333[$i]) −(($min+$max)/2)) + abs(abs($BN232[$i]) −(($min+$max)/2))*abs(
        abs($BN232[$i]) −(($min+$max)/2)) + abs(abs($BN233[$i]) −(($min+$max)/2))*abs
        (abs($BN233[$i]) −(($min+$max)/2)))/9/(($max−$min)/2)/(($max−$min)/2); #mean
         distance from frequencies to center. border areas are more relevant.
         maximum is 1. smaller is better.
     $n++;

     if (abs($BN111[$i]) > $max || abs($BN111[$i]) < $min || abs($BN122[$i]) > $max
         || abs($BN122[$i]) < $min || abs($BN121[$i]) > $max || abs($BN121[$i]) <
         $min || abs($BN133[$i]) > $max || abs($BN133[$i]) < $min || abs($BN131[$i])
          > $max || abs($BN131[$i]) < $min || abs($BN222[$i]) > $max || abs($BN222[
         $i]) < $min || abs($BN333[$i]) > $max || abs($BN333[$i]) < $min || abs(
         $BN232[$i]) > $max || abs($BN232[$i]) < $min || abs($BN233[$i]) > $max ||
         abs($BN233[$i]) < $min) {
71       last;
     }
  }
     return($i,$nice/$n); #divide by number of days

76 }
```

```
sub ausgabe {
    my $outstart = shift;
    my $outend = shift;
4   my $OutF = $filename.".txt";
    open(OUTPUT, ">>$OutF") or die "Can't open or create file $OutF: $!\n";
    my $item;
    for ($i=$outstart;$i<$outend;$i++) {
        print OUTPUT "$i ".$BN111[$i]." ".$BN122[$i]." ".$BN121[$i]." ".$BN133[$i]." ".
            $BN131[$i]." ".$BN222[$i]." ".$BN333[$i]." ".$BN232[$i]." ".$BN233[$i]."\n"
            ;
9   }
    close(OUTPUT);

    if ($i < $end) {
        my $BrkF = $filename.".brk";
14      open(BREAKS, ">>$BrkF") or die "Can't open or create file $OutF: $!\n";
        print BREAKS "set arrow from $i.5,$min to $i.5,$max nohead linewidth 2 front\n"
            ;
        close(BREAKS);
    }

19  return 1;
}
```

```
sub datum_zeit{

    my $p=$_[0];
    my %DATUM_ZEIT;
5   my $timeparameter;
```

```perl
        my ($Sekunden, $Minuten, $Stunden, $Monatstag, $Monat, $Jahr, $Wochentag,
            $Jahrestag, $Sommerzeit) = localtime(time);
        $Monat+=1;
        $Jahrestag+=1;
        $Monat = $Monat < 10 ? $Monat = "0".$Monat : $Monat;
        $Monatstag = $Monatstag < 10 ? $Monatstag = "0".$Monatstag : $Monatstag;
        $Stunden = $Stunden < 10 ? $Stunden = "0".$Stunden : $Stunden;
        $Minuten = $Minuten < 10 ? $Minuten = "0".$Minuten : $Minuten;
        $Sekunden = $Sekunden < 10 ? $Sekunden = "0".$Sekunden : $Sekunden;
        $Jahr+=1900;

        $DATUM_ZEIT{'J'}=$Jahr;
        $DATUM_ZEIT{'M'}=$Monat;
        $DATUM_ZEIT{'D'}=$Monatstag;
        $DATUM_ZEIT{'h'}=$Stunden;
        $DATUM_ZEIT{'m'}=$Minuten;
        $DATUM_ZEIT{'s'}=$Sekunden;
        $DATUM_ZEIT{'T'}=$Wochentag;


        if ($p){
            $p=~s/(J|M|D|h|m|s|T)/$DATUM_ZEIT{$1}/g, while ($p=~/J|M|D|h|m|s|T/);
        }
        else{
            $p=$Jahr.$Monat.$Monatstag.'_'.$Stunden.$Minuten.$ Sekunden;
        }
        return $p;
}
```

B.1   50 MHZ TTL BASED AMPLITUDE STABILIZATION

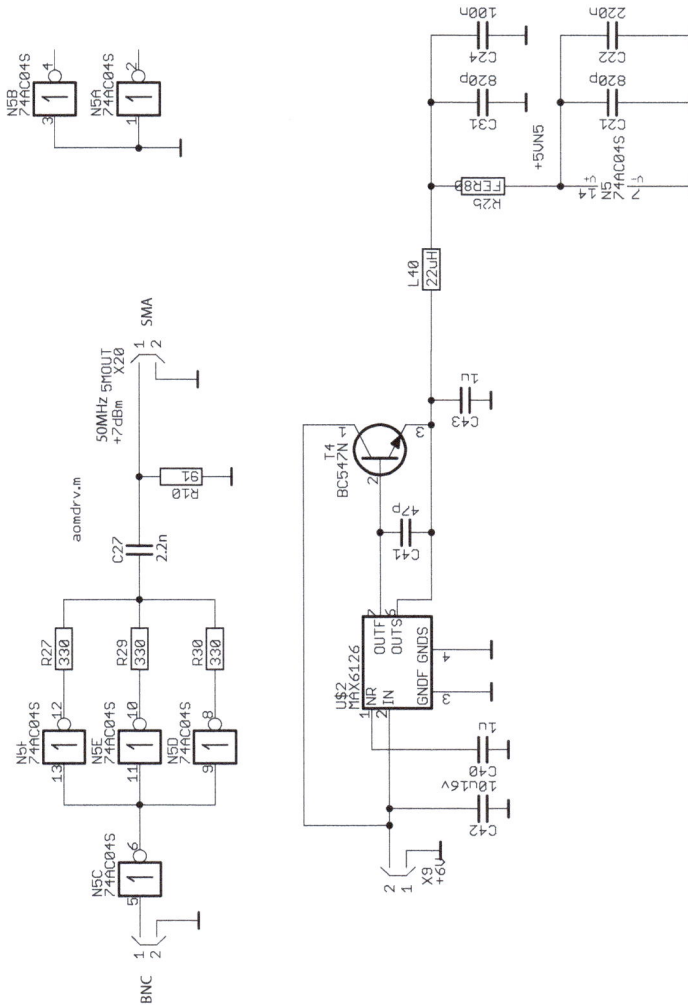

B.2 FDS BOARD (REVISION 1)

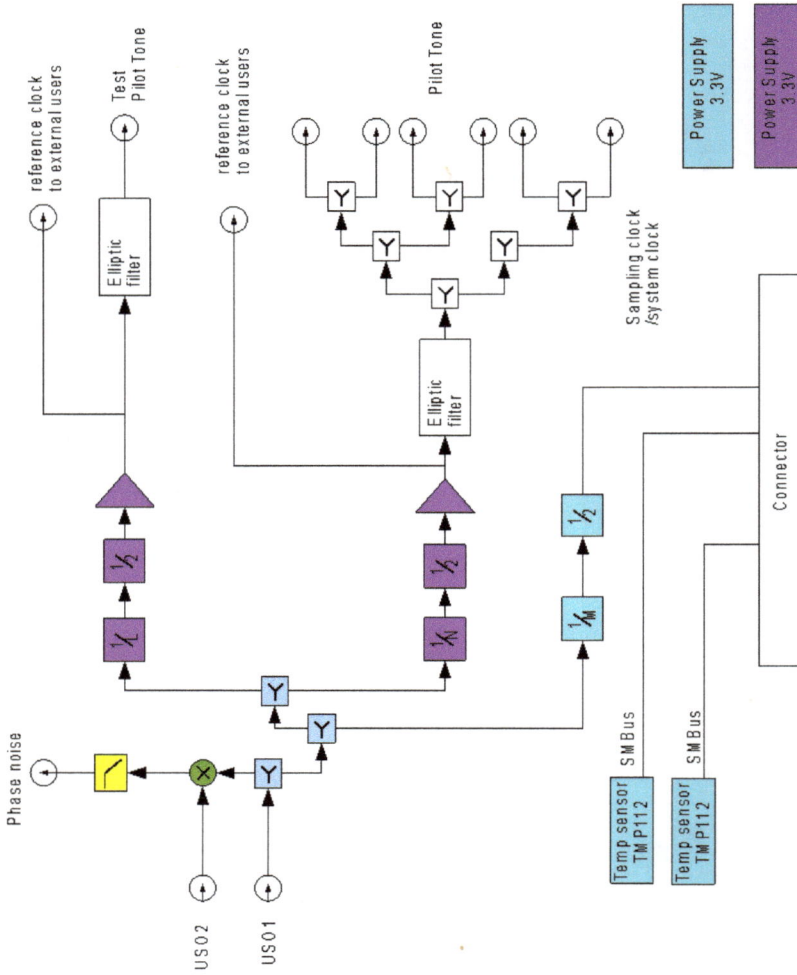

B.3 FDS BOARD (REVISION 2)

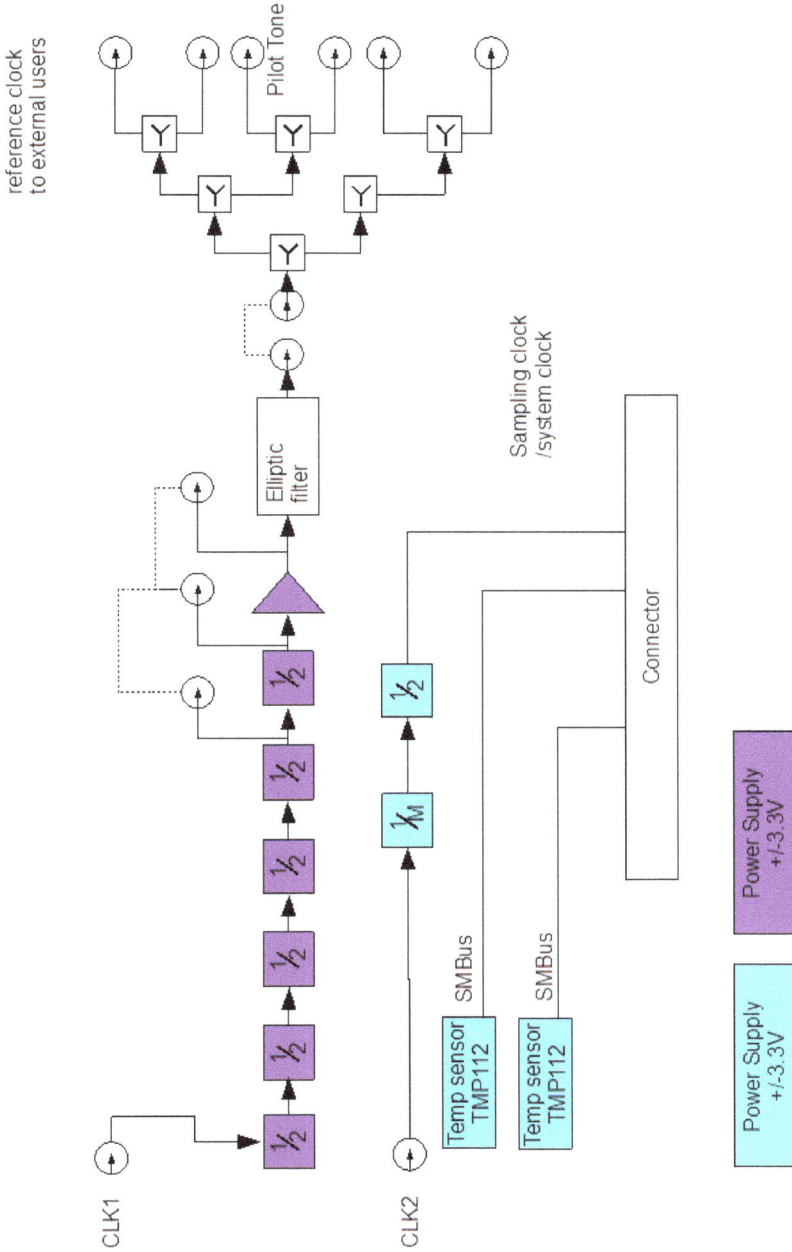

Part VI

ACRONYMS, LISTS & REFERENCES

AAF	Anti-Aliasing Filter
AC	Alternating Current
AEI	Albert Einstein Institute
AM CVn	AM Canum Venaticorum
ADC	Analog-to-Digital Converter
CDM	Cold Dark Matter
CGI	Common Gateway Interface
CMB	Cosmic Microwave Background
CV	Cataclysmic Variable
DAC	Digital-to-Analog Converter
DC	Direct Current
DFACS	Drag-Free and Attitude Control System
DPLL	Digital Phase-Locked Loop
DS/SS	Direct-Sequence Spread-Spectrum
DTU	Technical University of Denmark
eLISA	Evolving Laser Interferometer Space Antenna, refers to a constantly updated LISA mission concept
EMRI	Extreme Mass Ratio Inspiral
EOM	Electro-Optical Modulator
ESA	European Space Agency
ESTEC	European Space Research and Technology Centre
FA	Fiber Amplifier
FDS	Frequency Distribution System
GUI	Graphical User Interface
HTML	HyperText Markup Language
I	In-phase

IBEX	Interstellar Boundary Explorer
IT	Information Technology
ΛCDM	Lambda-CDM model
LIGO	Laser Interferometer Gravitational Wave Observatory
LISA	Laser Interferometer Space Antenna, refers either to the Classic LISA mission concept finalized in 2011 or the general type of LISA-like missions
LPSD	Linear Power Spectral Density
LUT	Look-Up Table
NASA	National Aeronautics and Space Administration
NCO	Numerically Controlled Oscillator
ND	Neutral Density
NGO	New Gravitational wave Observatory, refers to a LISA-like mission concept investigated by ESA
NPRO	Non-Planar Ring Oscillator
OGO	Octahedral Gravitational Observatory
OPD	Optical Path Difference
PA	Phase Accumulator
PBS	Polarizing Beam Splitter
PCB	Printed Circuit Board
PD	Photodetector
PDF	Portable Document Format
PDL	Perl Data Language
PE	Polyethylene
PI	Proportional-Integral (controller)
PID	Proportional-Integral-Differential (controller)
PIR	Phase Increment Register
PLL	Phase-Locked Loop
PMS	Phase Measurement System
PTFE	Polytetrafluoroethylene

Q	Quadrature phase
QUEST	Centre for Quantum Engineering and Space-Time Research
PRN	Pseudo-Random Noise
RIN	Relative Intensity Noise
RMS	Root Mean Square
S/C	Spacecraft
SAGA	a LISA-like mission concept with a specific set of parameters explored in this thesis
SMA	SubMiniature version A
SBB	Single Side-Band
SGO	Space-based Gravitational-wave Observatory, refers to a LISA-like mission concept investigated by NASA
SNR	Signal-to-Noise Ratio
SVG	Scalable Vector Graphics
TDI	Time Delay Interferometry
TIA	Trans-Impedance Amplifier
TRL	Technology Readiness Level
UF	University of Florida
UHF	Ultra High Frequency
USO	Ultra Stable Oscillator
VCO	Voltage Controlled Oscillator
VHDL	Very High speed integrated circuit Hardware Description Language
WD	White Dwarf
XOR	Exclusive-Or
ZARM	Center of Applied Space Technology and Microgravity

LIST OF FIGURES

LIST OF TABLES

[1] M.A. Finocchiaro. *The Galileo Affair: A Documentary History*. California studies in the history of science. University of California, 1989.

[2] A. Calaprice. *The Einstein Almanac*. Johns Hopkins University Press, 2005.

[3] DJ McComas, MA Dayeh, HO Funsten, G Livadiotis, and NA Schwadron. The heliotail revealed by the interstellar boundary explorer. *The Astrophysical Journal*, 771(2):77, 2013.

[4] T. J. Cox and A. Loeb. The collision between the Milky Way and Andromeda. 386:461–474, May 2008.

[5] Donald Lynden-Bell, SM Faber, David Burstein, Roger L Davies, Alan Dressler, RJ Terlevich, and Gary Wegner. Spectroscopy and photometry of elliptical galaxies. v-galaxy streaming toward the new supergalactic center. *The Astrophysical Journal*, 326:19–49, 1988.

[6] I. Newton. *Philosophiæ naturalis principia mathematica: 1687*. 1687.

[7] I. Newton and A. Janiak. *Isaac Newton: Philosophical Writings*. Cambridge Texts in the History of Philosophy. Cambridge University Press, 2004.

[8] J. Earman, M. Janssen, and J.D. Norton. *The Attraction of Gravitation: New Studies in the History of General Relativity*. Contemporary Mathematicians. Birkhäuser Boston, 1993.

[9] W.B. Taylor. *Kinetic Theories of Gravitation*. Number pp. 205-282 in Annual report. U.S. Government Printing Office, 1877.

[10] Albert Einstein. Die feldgleichungen der gravitation. *Albert Einstein: Akademie-Vorträge*, pages 88–92, 1915.

[11] J.A. Wheeler. *A Journey Into Gravity and Spacetime*. Scientific American Library paperback. Henry Holt and Company, 1999.

[12] Mario Livio. Lost in translation: Mystery of the missing text solved. *Nature*, 479(7372):171–173, 2011.

[13] Georges Lemaître. Un univers homogène de masse constante et de rayon croissant rendant compte de la vitesse radiale des nébuleuses extra-galactiques. *Annales de la Societe Scietifique de Bruxelles*, 47:49–59, 1927.

[14] Edwin Hubble. A relation between distance and radial velocity among extra-galactic nebulae. *Proceedings of the National Academy of Sciences*, 15(3):168–173, 1929.

[15] Adam G Riess, Alexei V Filippenko, Peter Challis, Alejandro Clocchiatti, Alan Diercks, Peter M Garnavich, Ron L Gilliland, Craig J Hogan, Saurabh Jha, Robert P Kirshner, et al. Observational evidence from supernovae for an accelerating universe and a cosmological constant. *The Astronomical Journal*, 116(3):1009, 1998.

[16] Aleksandr Friedmann. 125. on the curvature of space. *Zeitschrift für Physik*, 10:377–386, 1922.

[17] Ray A d'Inverno and Alex Harvey. Introducing einstein's relativity. *Physics Today*, 46(8):59–60, 2008.

[18] P James E Peebles. Dark matter and the origin of galaxies and globular star clusters. *The Astrophysical Journal*, 277:470–477, 1984.

[19] Massimo Persic, Paolo Salucci, and Fulvio Stel. The universal rotation curve of spiral galaxies—i. the dark matter connection. *Monthly Notices of the Royal Astronomical Society*, 281(1):27–47, 1996.

[20] Marc Davis, George Efstathiou, Carlos S Frenk, and Simon DM White. The evolution of large-scale structure in a universe dominated by cold dark matter. *The Astrophysical Journal*, 292:371–394, 1985.

[21] N Jarosik, CL Bennett, J Dunkley, B Gold, MR Greason, M Halpern, RS Hill, G Hinshaw, A Kogut, E Komatsu, et al. Seven-year wilkinson microwave anisotropy probe (wmap) observations: sky maps, systematic errors, and basic results. *The Astrophysical Journal Supplement Series*, 192(2):14, 2011.

[22] Robert R Caldwell, Marc Kamionkowski, and Nevin N Weinberg. Phantom energy: dark energy with w<-1 causes a cosmic doomsday. *Physical Review Letters*, 91(7):071301, 2003.

[23] L.M. Krauss. *A Universe from Nothing: Why There Is Something Rather Than Nothing*. Atria Books, 2012.

[24] Laura Mersini-Houghton and Harald P Pfeiffer. Back-reaction of the hawking radiation flux on a gravitationally collapsing star ii: Fireworks instead of firewalls. *arXiv preprint arXiv:1409.1837*, 2014.

[25] Stephen W Hawking. Black hole explosions. *Nature*, 248(5443):30–31, 1974.

[26] Don N Page. Information in black hole radiation. *Physical review letters*, 71(23):3743, 1993.

[27] Ahmed Almheiri, Donald Marolf, Joseph Polchinski, and James Sully. Black holes: complementarity or firewalls? *Journal of High Energy Physics*, 2013(2):1–20, 2013.

[28] J Richard Gott III, Mario Jurić, David Schlegel, Fiona Hoyle, Michael Vogeley, Max Tegmark, Neta Bahcall, and Jon Brinkmann. A map of the universe. *The Astrophysical Journal*, 624(2):463, 2005.

[29] Hugh Everett. The theory of the universal wave function. 1973.

[30] Alan H Guth. Inflationary universe: A possible solution to the horizon and flatness problems. *Physical Review D*, 23(2):347, 1981.

[31] Paul J Steinhardt and Neil Turok. *Endless universe: Beyond the big bang*. Broadway, 2007.

[32] Justin Khoury, Burt A Ovrut, Paul J Steinhardt, and Neil Turok. Ekpyrotic universe: Colliding branes and the origin of the hot big bang. *Physical Review D*, 64(12):123522, 2001.

[33] Albert Einstein and Nathan Rosen. On gravitational waves. *Journal of the Franklin Institute*, 223(1):43–54, 1937.

[34] E.F. Taylor, J.A. Wheeler, and E.W. Bertschinger. *Exploring Black Holes: Introduction to General Relativity*. Addison-Wesley, 2010.

[35] Joseph H Taylor and Joel M Weisberg. Further experimental tests of relativistic gravity using the binary pulsar PSR 1913+ 16. *The Astrophysical Journal*, 345:434–450, 1989.

[36] PA Ade, RW Aikin, D Barkats, SJ Benton, CA Bischoff, JJ Bock, JA Brevik, I Buder, E Bullock, CD Dowell, et al. Detection of B-Mode Polarization at Degree Angular Scales by BICEP2. *Phys. Rev. Lett.*, 112:241101, Jun 2014.

[37] Michael J. Mortonson and Uroš Seljak. A joint analysis of Planck and BICEP2 B modes including dust polarization uncertainty. *Journal of Cosmology and Astroparticle Physics*, 2014(10):035, 2014.

[38] Christian D. Ott, Adam Burrows, Eli Livne, and Rolf Walder. Gravitational waves from axisymmetric, rotating stellar core collapse. *The Astrophysical Journal*, 600(2):834, 2004.

[39] Chris L. Fryer and Kimberly C.B. New. Gravitational waves from gravitational collapse. *Living Reviews in Relativity*, 6(2), 2003.

[40] R. A. Hulse and J. H. Taylor. Discovery of a pulsar in a binary system. *Astrophysical Journal, Letters*, 195:L51–L53, January 1975.

[41] S. Sigurdsson, J. Wright, R. Griffith, and M. S. Povich. A Mid-Infrared Search for Kardashev Civilizations. In *American Astronomical Society Meeting Abstracts*, volume 223 of *American Astronomical Society Meeting Abstracts*, page 349.01, January 2014.

[42] Alexander Vilenkin. Cosmic strings and domain walls. *Physics Reports*, 121(5):263–315, 1985.

[43] N. S. Kardashev. Transmission of Information by Extraterrestrial Civilizations. *Soviet Astronomy*, 8:217, October 1964.

[44] G. H. Janssen, B. W. Stappers, M. Kramer, M. Purver, A. Jessner, and I. Cognard. European pulsar timing array. *AIP Conference Proceedings*, 983(1):633–635, 2008.

[45] G Hobbs, A Archibald, Z Arzoumanian, D Backer, M Bailes, N D R Bhat, M Burgay, S Burke-Spolaor, D Champion, I Cognard, W Coles, J Cordes, P Demorest, G Desvignes, R D Ferdman, L Finn, P Freire, M Gonzalez, J Hessels, A Hotan, G Janssen, F Jenet, A Jessner, C Jordan, V Kaspi, M Kramer, V Kondratiev, J Lazio, K Lazaridis, K J Lee, Y Levin, A Lommen, D Lorimer, R Lynch, A Lyne, R Manchester, M McLaughlin, D Nice, S Oslowski, M Pilia, A Possenti, M Purver, S Ransom, J Reynolds, S Sanidas, J Sarkissian, A Sesana, R Shannon, X Siemens, I Stairs, B Stappers, D Stinebring, G Theureau, R van Haasteren, W van Straten, J P W Verbiest, D R B Yardley, and X P You. The International Pulsar Timing Array project: using pulsars as a gravitational wave detector. *Classical and Quantum Gravity*, 27(8):084013, 2010.

[46] J. Weber. Detection and generation of gravitational waves. *Phys. Rev.*, 117:306–313, Jan 1960.

[47] Gregory M Harry, LIGO Scientific Collaboration, et al. Advanced LIGO: the next generation of gravitational wave detectors. *Classical and Quantum Gravity*, 27(8):084006, 2010.

[48] T Accadia, F Acernese, F Antonucci, P Astone, G Ballardin, F Barone, M Barsuglia, A Basti, Th S Bauer, M Bebronne, et al. Status of the Virgo project. *Classical and Quantum Gravity*, 28(11):114002, 2011.

[49] Kentaro Somiya. Detector configuration of KAGRA–the Japanese cryogenic gravitational-wave detector. *Classical and Quantum Gravity*, 29(12):124007, 2012.

[50] Hartmut Grote, LIGO Scientific Collaboration, et al. The status of GEO 600. *Classical and Quantum Gravity*, 25(11):114043, 2008.

[51] P Amaro Seoane, Sofiane Aoudia, H Audley, G Auger, S Babak, J Baker, E Barausse, S Barke, M Bassan, V Beckmann, et al. The Gravitational Universe. *arXiv preprint arXiv:1305.5720*, 2013.

[52] The European Space Agency. ESA's new vision to study the invisible universe. Website: `http://www.esa.int/Our_Activities/Space_Science/ESA_s_new_vision_to_study_the_invisible_Universe`, 11 2013. accessed 2015-01-12.

[53] eLISA Consortium. The Gravitational Universe. `https://support.elisascience.org/`. Accessed: 2015-01-18.

[54] Antoine Petiteau. private communication, 2013.

[55] Richard Powell. An Atlas of the Universe. Website: `www.atlasoftheuniverse.com/superc.html`, 2006. accessed 2015-01-12.

[56] Alberto Sesana. private communication, 2013.

[57] Elliott P Horch, Steve B Howell, Mark E Everett, and David R Ciardi. Most Sub-Arcsecond Companions of Kepler Exoplanet Candidate Host Stars are Gravitationally Bound. *arXiv preprint arXiv:1409.1249*, 2014.

[58] Gijs Nelemans. LISA wiki (verification binaries). `http://www.astro.ru.nl/~nelemans/dokuwiki/doku.php?id=lisa_wiki`. Accessed: 2014-09-01.

[59] Cole Miller. Lecture 25: Grav waves from binaries. `http://www.astro.umd.edu/~miller/teaching/astr498/`, 2008. accessed 2015-01-12.

[60] Simon Barke, Yang Wang, Juan Jose Esteban Delgado, Michael Tröbs, Gerhard Heinzel, and Karsten Danzmann. Towards a Gravitational Wave Observatory Designer: Sensitivity Limits of Space-borne Detectors. *Classical and Quantum Gravity*, 32(9):095004, 2015.

[61] A. Orchard. *Cassell's Dictionary of Norse Myth and Legend.* Cassell Dictionary of ... Series. Cassell, 2002.

[62] Yan Wang, David Keitel, Stanislav Babak, Antoine Petiteau, Markus Otto, Simon Barke, Fumiko Kawazoe, Alexander Khalaidovski, Vitali Müller, Daniel Schütze, et al. Octahedron configuration for a displacement noise-cancelling gravitational wave detector in space. *Physical Review D*, 88(10):104021, 2013.

[63] S Phinney, P Bender, R Buchman, R Byer, N Cornish, P Fritschel, W Folkner, S Merkowitz, K Danzmann, L DiFiore, et al. The big bang observer: direct detection of gravitational waves from the birth of the universe to the present. *NASA mission concept study*, 2004.

[64] DA Holmes, PV Avizonis, and KH Wrolstad. On-axis irradiance of a focused, apertured Gaussian beam. *Applied optics*, 9(9):2179–2180, 1970.

[65] Gerhard Heinzel, Juan José Esteban, Simon Barke, Markus Otto, Yan Wang, Antonio F Garcia, and Karsten Danzmann. Auxiliary functions of the LISA laser link: ranging, clock noise transfer and data communication. *Classical and Quantum Gravity*, 28(9):094008, 2011.

[66] M Tröbs, S Barke, J Möbius, M Engelbrecht, D Kracht, L d'Arcio, G Heinzel, and K Danzmann. Lasers for LISA: Overview and phase characteristics. In *Journal of Physics: Conference Series*, volume 154, page 012016. IOP Publishing, 2009.

[67] Guido Mueller, Paul McNamara, Ira Thorpe, and Jordan Camp. Laser frequency stabilization for lisa. Technical report, NASA, 2005. TM-2005-212794.

[68] Volker Leonhardt and Jordan B Camp. Space interferometry application of laser frequency stabilization with molecular iodine. *Applied optics*, 45(17):4142–4146, 2006.

[69] E J Elliffe, J Bogenstahl, A Deshpande, J Hough, C Killow, S Reid, D Robertson, S Rowan, H Ward, and G Cagnoli. Hydroxide-catalysis bonding for stable optical systems for space. *Classical and Quantum Gravity*, 22(10):S257, 2005.

[70] O Jennrich. LISA technology and instrumentation. *Classical and Quantum Gravity*, 26(15):153001, 2009.

[71] F Guzman Cervantes, J Livas, R Silverberg, E Buchanan, and R Stebbins. Characterization of photoreceivers for lisa. *Classical and Quantum Gravity*, 28(9):094010, 2011.

[72] Brian J Meers and Kenneth A Strain. Modulation, signal, and quantum noise in interferometers. *Physical Review A*, 44(7):4693, 1991.

[73] TM Niebauer, Roland Schilling, Karsten Danzmann, Albrecht Rüdiger, and Walter Winkler. Nonstationary shot noise and its effect on the sensitivity of interferometers. *Physical Review A*, 43(9):5022, 1991.

[74] Simon Barke, Nils Brause, Iouri Bykov, Juan Jose Esteban Delgado, Anders Enggaard, Oliver Gerberding, Gerhard Heinzel, Joachim Kullmann, Søren Møller Pedersen, and Torben Rasmussen. *LISA Metrology Systen - Final Report*. DTU Space / AEI Hannover / Axcon ApS, 2014.

[75] Juan José Esteban, Antonio F. García, Simon Barke, Antonio M. Peinado, Felipe Guzmán Cervantes, Iouri Bykov, Gerhard Heinzel, and Karsten Danzmann. Experimental demonstration of weak-light laser ranging and data communication for LISA. *Optics Express*, 19(17):15937–15946, 2011.

[76] Simon Barke. Inter-Spacecraft Clock Transfer Phase Stability for LISA. Diploma thesis, Leibniz Universität Hannover, Institute for Gravitational Physics, 2008.

[77] Alix Preston. *Stability of materials for use in space-based interferometric missions*. 2010.

[78] Sönke Schuster, Gudrun Wanner, Michael Tröbs, and Gerhard Heinzel. Vanishing tilt-to-length coupling for a singular case in two-beam laser interferometers with Gaussian beams. *arXiv preprint arXiv:1406.5367*, 2014. accepted for publication in Applied Optics.

[79] Ke-Xun Sun, Ulrich Johann, Dan B DeBra1, Sasha Buchman, and Robert L Byer. Lisa gravitational reference sensors. *Journal of Physics: Conference Series*, 60(1):272, 2007.

[80] A Cavalleri, G Ciani, R Dolesi, M Hueller, D Nicolodi, D Tombolato, P J Wass, W J Weber, S Vitale, and L Carbone. Direct force measurements for testing the lisa pathfinder gravitational reference sensor. *Classical and Quantum Gravity*, 26(9):094012, 2009.

[81] T J Sumner, D N A Shaul, M O Schulte, S Waschke, D Hollington, and H Araújo. Lisa and lisa pathfinder charging. *Classical and Quantum Gravity*, 26(9):094006, 2009.

[82] S. E. Pollack, M. D. Turner, S. Schlamminger, C. A. Hagedorn, and J. H. Gundlach. Charge management for gravitational-wave observatories using uv leds. *Phys. Rev. D*, 81:021101, Jan 2010.

[83] T. Ziegler, P. Bergner, G. Hechenblaikner, N. Brandt, and W. Fichter. Modeling and performance of contact-free discharge systems for space inertial sensors. *IEEE Transactions on Aerospace Electronic Systems*, 50:1493–1510, April 2014.

[84] F Antonucci, M Armano, H Audley, G Auger, M Benedetti, P Binetruy, C Boatella, J Bogenstahl, D Bortoluzzi, P Bosetti, N Brandt, M Caleno, A Cavalleri, M Cesa, M Chmeissani, G Ciani, A Conchillo, G Congedo, I Cristofolini, M Cruise, K Danzmann, F De Marchi, M Diaz-Aguilo, I Diepholz, G Dixon, R Dolesi, N Dunbar, J Fauste, L Ferraioli, D Fertin, W Fichter, E Fitzsimons, M Freschi, A García Marin, C García Marirrodriga, R Gerndt, L Gesa, D Giardini, F Gibert, C Grimani, A Grynagier, B Guillaume, F Guzmán, I Harrison, G Heinzel, M Hewitson, D Hollington, J Hough, D Hoyland, M Hueller, J Huesler, O Jeannin, O Jennrich, P Jetzer, B Johlander, C Killow, X Llamas, I Lloro, A Lobo, R Maarschalkerweerd, S Madden, D Mance, I Mateos, P W McNamara, J Mendes, E Mitchell, A Monsky, D Nicolini, D Nicolodi, M Nofrarias, F Pedersen, M Perreur-Lloyd, A Perreca, E Plagnol, P Prat, G D Racca, B Rais, J Ramos-Castro, J Reiche, J A Romera Perez, D Robertson, H Rozemeijer, J Sanjuan, A Schleicher, M Schulte, D Shaul, L Stagnaro, S Strandmoe, F Steier, T J Sumner, A Taylor, D Texier, C Trenkel, D Tombolato, S Vitale, G Wanner, H Ward, S Waschke, P Wass, W J Weber, and P Zweifel. From laboratory experiments to lisa pathfinder: achieving lisa geodesic motion. *Classical and Quantum Gravity*, 28(9):094002, 2011.

[85] P McNamara, S Vitale, K Danzmann, and on behalf of the LISA Pathfinder Science Working Team. Lisa pathfinder. *Classical and Quantum Gravity*, 25(11):114034, 2008.

[86] S Anza, M Armano, E Balaguer, M Benedetti, C Boatella, P Bosetti, D Bortoluzzi, N Brandt, C Braxmaier, M Caldwell, et al. The LTP experiment on the LISA Pathfinder mission. *Classical and Quantum Gravity*, 22(10):S125, 2005.

[87] N. Brandt. LISA Technology Package Experiment Performance Budget: Issue 2.4, 2014. unpublished.

[88] Markus Otto, Gerhard Heinzel, and Karsten Danzmann. TDI and clock noise removal for the split interferometry configuration of LISA. *Classical and Quantum Gravity*, 29(20):205003, 2012.

[89] Massimo Tinto and Sanjeev V Dhurandhar. Time-Delay Interferometry. *Living Reviews in Relativity*, 17(6), 2014.

[90] Juan José Esteban, Iouri Bykov, Antonio Francisco García Marín, Gerhard Heinzel, and Karsten Danzmann. Optical ranging and data transfer development for LISA. In *Journal of Physics: Conference Series*, volume 154, page 012025. IOP Publishing, 2009.

[91] Yan Wang, Gerhard Heinzel, and Karsten Danzmann. First stage of LISA data processing: Clock synchronization and arm-length determination via a hybrid-extended kalman filter. *Phys. Rev. D*, 90:064016, Sep 2014.

[92] Yan Wang. *On inter-satellite laser ranging, clock synchronization and gravitational wave data analysis*. PhD thesis, Leibniz Universität Hannover, Institute for Gravitational Physics, 2014.

[93] Roland Schilling. Angular and frequency response of lisa. *Classical and Quantum Gravity*, 14(6):1513, 1997.

[94] Walter J Doherty and Arvind J Thadhani. The economic value of rapid response time. *IBM Report GE20-0752-0*, 11 1982.

[95] Massimo Tinto, Frank B Estabrook, and JW Armstrong. Time-delay interferometry for LISA. *Physical Review D*, 65(8):082003, 2002.

[96] Massimo Tinto and Sanjeev V. Dhurandhar. Time-Delay Interferometry. *Living Reviews in Relativity*, 17(6), 2014.

[97] Neil J Cornish. Detecting a stochastic gravitational wave background with the Laser Interferometer Space Antenna. *Physical Review D*, 65(2):022004, 2001.

[98] Yanbei Chen, Archana Pai, Kentaro Somiya, Seiji Kawamura, Shuichi Sato, Keiko Kokeyama, Robert L Ward, Keisuke Goda, and Eugeniy E

Mikhailov. Interferometers for displacement-noise-free gravitational-wave detection. *Physical review letters*, 97(15):151103, 2006.

[99] O Jenrich, eLISA/NGO Collaboration, et al. NGO Assessment Study Report (Yellow Book) ESA. *SRE (2011)*, 19, 2012.

[100] K. Danzmann, T. A. Prince, P. Binetruy, P. Bender, S. Buchman, J. Centrella, M. Cerdonio, N. Cornish, M. Cruise, C. J. Cutler, L. S. Finn, J. Gundlach, C. Hogan, J. Hough, S. A. Hughes, O. Jennrich, P. Jetzer, A. Lobo, P. Madau, Y. Mellier, S. Phinney, D. O. Richstone, B. Schutz, R. Stebbins, T. Sumner, K. Thorne, J.-Y. Vinet, and S. Vitale. LISA Assessment Study Report. 2011.

[101] The European Space Agency. Timeline for Selection of L-Class Missions. Website: `http://sci.esa.int/cosmic-vision/42369-l-class-timeline/`, 2014. accessed 2015-01-12.

[102] K Danzmann, T Prince, et al. LISA assessment study report (Yellow Book). Technical report, 2011.

[103] LISA Study Team. Mission formulation. Technical report, EADS Astrium, Germany, 2006.

[104] Patrick Kwee, Benno Willke, and Karsten Danzmann. Laser power noise detection at the quantum-noise limit of 32?a photocurrent. *Opt. Lett.*, 36(18):3563–3565, Sep 2011.

[105] Patrick Kwee. Diagnostic breadboard instruction manual. Technical report, Albert Einstein Institute, June 2009.

[106] Michael Tröbs, Peter Wessels, and Carsten Fallnich. Phase-noise properties of an ytterbium-doped fiber amplifier for the laser interferometer space antenna. *Optics letters*, 30(7):789–791, 2005.

[107] Michael Tröbs, Carsten Fallnich, et al. Power-and frequency-noise characteristics of an yb-doped fiber amplifier and actuators for stabilization. *Optics express*, 13(6):2224–2235, 2005.

[108] Oliver Jennrich. private communication, 2014.

[109] Hans-Georg Beyer and Bernhard Sendhoff. Robust optimization–a comprehensive survey. *Computer methods in applied mechanics and engineering*, 196(33):3190–3218, 2007.

[110] Alf Tang and Timothy J. Sumner. Removing the trend of drift induced from acceleration noise for LISA. 2012.

[111] Massimo Tinto and John W. Armstrong. Cancellation of laser noise in an unequal-arm interferometer detector of gravitational radiation. *Physical Review D: Particles and Fields*, 59:102003, 1999.

[112] Markus Otto, Gerhard Heinzel, and Karsten Danzmann. Tdi and clock noise removal for the split interferometry configuration of lisa. *Class. Quantum Grav.*, 29:205003, 2012.

[113] Juan José Esteban Delgado. *Laser Raging and Data Communication for the Laser Inerferometer Space Antenna*. PhD thesis, Gottfried Wilhelm Leibniz Universität Hannover, 2012.

[114] Stephen B Wicker and Vijay K Bhargava. *Reed-Solomon codes and their applications*. Wiley. com, 1999.

[115] LISA Metrology System Team. Metrology system design report. Technical Report TN2-1:AO/1-6238/10/NL/HB, European Space Agency, 2012.

[116] Daniel Edler. Measurement and Modelling of USO Clock Noise in space based applications. Bachelor thesis, Leibniz Universität Hannover, Institute for Gravitational Physics, 2014.

[117] T.E. Parker. Characteristics and sources of phase noise in stable oscillators. In *41st Annual Symposium on Frequency Control. 1987*, pages 99–110, May 1987.

[118] Seth M. Foreman, Adela Marian, Jun Ye, Evgeny A. Petrukhin, Mikhail A. Gubin, Oliver D. Mücke, Franco N. C. Wong, Erich P. Ippen, and Franz X. Kärtner. Demonstration of a hene/ch4-based opticalmolecular clock. *Opt. Lett.*, 30(5):570–572, Mar 2005.

[119] John G Hartnett, Nitin R Nand, and Chuan Lu. Ultra-low-phase-noise cryocooled microwave dielectric-sapphire-resonator oscillators. *Applied Physics Letters*, 100(18):183501, 2012.

[120] Joachim Kullmann. *Development of a digital phase measuring system with microradian precision for LISA*. PhD thesis, Leibniz Universität Hannover, 2012.

[121] Simon Barke, Michael Trös, Benjamin Sheard, Gerhard Heinzel, and Karsten Danzmann. Phase noise contribution of EOMs and HF cables. *Journal of Physics: Conference Series*, 154(1):012006, 2009.

[122] M Tröbs, S Barke, J Möbius, M Engelbrecht, Th Theeg, D Kracht, B Sheard, G Heinzel, and K Danzmann. Fiber modulators and fiber amplifiers for LISA. In *Journal of Physics: Conference Series*, volume 228, page 012042. IOP Publishing, 2010.

[123] M Tröbs, S Barke, Th Theeg, D Kracht, G Heinzel, and K Danzmann. Differential phase-noise properties of a ytterbium-doped fiber amplifier for the Laser Interferometer Space Antenna. *Optics letters*, 35(3):435–437, 2010.

[124] S Barke, M Tröbs, B Sheard, G Heinzel, and K Danzmann. EOM sideband phase characteristics for the spaceborne gravitational wave detector LISA. *Applied Physics B*, 98(1):33–39, 2010.

[125] Satish K Dhawan. Understanding effect of teflon room temperature phase transition on coax cable delay in order to improve the measurement of te signals of deuterated polarized targets. *Nuclear Science, IEEE Transactions on*, 39(5):1331–1335, 1992.

[126] K Czuba and D Sikora. Temperature stability of coaxial cables. *Acta Phys. Pol. A*, 119(EuCARD-PUB-2011-001):553, 2011.

[127] Amrit Pal Singh. Computation of phase-stability over temperature - Thermal testing of microwave cables for the Laser Interferometer Space Antenna. Bachelor thesis, Leibniz Universität Hannover, Institute for Gravitational Physics, 2010.

[128] Gerhard Hejc, W Schafer, Achim Seidel, M-P Hess, Johannes Kehrer, and Giorgio Santarelli. Frequency dependency of phase stability of rf cables. In *Frequency Control Symposium, 2007 Joint with the 21st European Frequency and Time Forum. IEEE International*, pages 1309–1311. IEEE, 2007.

[129] Gerhard Heinzel. Advanced optical techniques for laser-interferometric gravitational-wave detectors. 1999.

[130] Juan José Esteban Delgado. *Laser ranging and data communication for the Laser Interferometer Space Antenna*. PhD thesis, Universidad de Granada, 2012.

[131] Mitsuhiro Tateda, Shigeru Tanaka, and Yasuyuki Sugawara. Thermal characteristics of phase shift in jacketed optical fibers. *Applied optics*, 19(5):770–773, 1980.

[132] Oliver Gerberding. *Phase readout for satellite interferometry*. PhD thesis, Leibniz Universität Hannover, 2014.

[133] Marina Dehne. *Construction and noise behaviour of ultra-stable optical systems for space interferometers*. PhD thesis, Leibniz Universität Hannover, 2012.

[134] Shawn J Mitryk. *Laser noise mitigation through time delay interferometry for space-based gravitational wave interferometers using the UF laser interferometry simulator*. 2012.

[135] Ulrich Velte. *Orbit simulations and optical phase locking techniques for an atom interferometric test of the universality of free fall*. PhD thesis, Leibniz Universität Hannover, Institute for Quantum Optics, 2015.

[136] LISA Metrology System Team. Metrology system BB requirements specification. Technical Report TN1-1:AO/1-6238/10/NL/HB, European Space Agency, 2011.

[137] LISA Metrology System Team. Metrology system consolidated architecture. Technical Report TN1-2:AO/1-6238/10/NL/HB, European Space Agency, 2011.

[138] LISA Metrology System Team. Metrology system development model design description and test results. Technical Report TN1-3:AO/1-6238/10/NL/HB, European Space Agency, 2011.

[139] LISA Metrology System Team. Metrology system test plan. Technical Report TN1-4:AO/1-6238/10/NL/HB, European Space Agency, 2011.

[140] LISA Metrology System Team. Test set-up design report. Technical Report TN1-5:AO/1-6238/10/NL/HB, European Space Agency, 2011.

[141] LISA Metrology System Team. Manufacturing Drawings. Technical Report TN2-2:AO/1-6238/10/NL/HB, European Space Agency, 2012.

[142] LISA Metrology System Team. Metrology system as built design report. Technical Report TN3-1:AO/1-6238/10/NL/HB, European Space Agency, 2012.

[143] LISA Metrology System Team. Test Set-up commissioning report. Technical Report TN3-2:AO/1-6238/10/NL/HB, European Space Agency, 2012.

[144] LISA Metrology System Team. As Built Configuration List. Technical Report TN3-3:AO/1-6238/10/NL/HB, European Space Agency, 2012.

[145] LISA Metrology System Team. Metrology system test report. Technical Report TN4-1:AO/1-6238/10/NL/HB, European Space Agency, 2013.

[146] LISA Metrology System Team. Lessons learnt summary report. Technical Report TN5-1:AO/1-6238/10/NL/HB, European Space Agency, 2014.

[147] LISA Metrology System Team. Metrology system and GSE User Manual. Technical Report TN5-2:AO/1-6238/10/NL/HB, European Space Agency, 2014.

[148] LISA Metrology System Team. Road map for Metrology system Development up to Flight Model (and ROM quotation). Technical Report TN5-3:AO/1-6238/10/NL/HB, European Space Agency, 2014.

[149] A. Sesana, W.J. Weber, C.J. Killow, M. Perreur-Lloyd, D.I. Robertson, H. Ward, E.D. Fitzsimons, J. Bryant, A.M. Cruise, G. Dixon, D. Hoyland, D. Smith, J. Bogenstahl, P.W. McNamara, R. Gerndt, R. Flatscher, G. Hechenblaikner, M. Hewitson, O. Gerberding, S. Barke, N. Brause, I. Bykov, K. Danzmann, A. Enggaard, A. Gianolio, T. Vendt Hansen, G. Heinzel, A. Hornstrup, O. Jennrich, J. Kullmann, S. Møller-Pedersen, T. Rasmussen, J. Reiche, Z. Sodnik, M. Suess, M. Armano, T. Sumner,

P.L. Bender, T. Akutsu, and B.S. Sathyaprakash. Space-based detectors. *General Relativity and Gravitation*, 46(12), 2014.

[150] M Hendry, C Bradaschia, H Audley, S Barke, DG Blair, N Christensen, K Danzmann, A Freise, O Gerberding, B Knispel, et al. Education and public outreach on gravitational-wave astronomy. *General Relativity and Gravitation*, 46(8):1–11, 2014.

[151] O. Gerberding, S. Barke, I. Bykov, K. Danzmann, A. Enddaard, J. J. Esteban, A. Gianolio, T. V. Hansen, G. Heinzel, A. Hornstrup, O. Jennrich, J. Kullmann, S. M. Pedersen, T. Rasmussen, Z. Sodnik, and M. Suess. Breadboard model of the LISA phasemeter. In G. Auger, P. Binetruy, and E. Plagnol, editors, *ASP Conference Series; 9th LISA Symposium*, volume 467 of *ASP Conference Series*, pages 271 – 275, 2012.

[152] M Tröbs, S Barke, J Möbius, M Engelbrecht, D Kracht, L d'Arcio, G Heinzel, and K Danzmann. Lasers for lisa: Overview and phase characteristics. In *Journal of Physics: Conference Series*, volume 154, page 012016. IOP Publishing, 2009.

[153] Sebastian Steinlechner, Simon Barke, Jessica Dück, Lars Hoppe, Raoul Amadeus Lorbeer, Markus Otto, Aiko Samblowski, and Tobias Westphal. Der lifter-ein flugobjekt mit ionenantrieb. *PhyDid A-Physik und Didaktik in Schule und Hochschule*, 2(7):20–26, 2008.

[154] Michael Tröbs and Gerhard Heinzel. Improved spectrum estimation from digitized time series on a logarithmic frequency axis. *Measurement*, 39(2):120–129, 2006.

[155] Michael Tröbs and Gerhard Heinzel. Improved spectrum estimation from digitized time series on a logarithmic frequency axis (vol 39, pg 120, 2006). *Measurement*, 42(1):170–170, 2009.

ABOUT THE AUTHOR

Simon Barke was born in Germany where he studied physics at the Leibniz Universität Hannover. For a while, he lived in Moshi, Tanzania, and worked at different schools and universities. Simon has always been curious and fascinated about physics and technology. Over the last years, he had the privilege to be involved in a number of exciting projects. This includes many talks at international conferences, a series of lectures held at various universities in China, different social media and public relations projects, exhibitions, shows, general interest talks, and ceremonial addresses. His research on low-frequency gravitational wave observatories in space was conducted at the University of Florida (Gainesville, FL, US) and the Albert Einstein Institute (Hanover, Germany) with support from the Centre for Quantum Engineering and Space-Time Research (QUEST) and the European Space Agency (ESA).

Address
Simon Barke
Dachenhausenstr. 5
D-30169 Hannover
Germany

Born October 22nd, 1980, in Hanover, Germany

Email mail@simonbarke.com

Website www.simonbarke.com

Research associate   since 2015
Max Planck Institute for Gravitational Physics, Hannover, Germany: laser interferometry in space in the group of PD Dr. Gerhard Heinzel.

Member of the LISA Metrology System Team   2011–2014
Max Planck Institute for Gravitational Physics, Hannover, Germany: design, development, manufacturing, testing and validation of a breadboard model of the LISA Metrology System under ESA contract.

Workshops   2004, 2012, 2013
Universities and secondary schools in Tanzania: preparation and supervision of workshops in astronomy and experimental physics.

Scientific monitor   2010
LIGO Livingston, LA, US: monitoring of the final Science Run (S6) of Enhanced LIGO, a 4 km ground-based gravitational wave detector.

Series of lectures   2009
Various universities, China: talks on gravitational wave astronomy in Shanghai, Wuhan, Nanjing, and Beijing.

Post-graduate student   2009–2014
Centre for Quantum Engineering and Space-Time Research (QUEST), Hanover, Germany: research on future spaceborne gravitational wave observatories.

Exchange researcher   2008-2009
University of Florida, Gainesville, FL, US: research on timing stability of fractional-N synthesizers in the group of Prof. Dr. Guido Müller.

Teacher   2004
Vunjo Secondary School, Moshi, Tanzania: three-months external internship, theoretical and experimental physics, mathematics.

Graduate physicist   2001-2008
Leibniz Universität Hannover, Germany: Dipl.-Phys., MSc equivalent.

Research assistant   2001–2004
Leibniz Universität Hannover, Germany: improvement of experiments, development of new experiments, teaching.

Community service   2000-2001
AWO-Seniorenzentrum Lehrte, Germany: support for assisted living.

High-school   1993-2000
Gymnasium Lehrte, Germany.

◆ Simon Barke, Yang Wang, Juan Jose Esteban Delgado, Michael Tröbs, Gerhard Heinzel, and Karsten Danzmann. Towards a Gravitational Wave Observatory Designer: Sensitivity Limits of Spaceborne Detectors. *Classical and Quantum Gravity*, 32(9):095004, 2015.

◆ Simon Barke, Nils Brause, Iouri Bykov, Juan Jose Esteban Delgado, Anders Enggaard, Oliver Gerberding, Gerhard Heinzel, Joachim Kullmann, Søren Møller Pedersen, and Torben Rasmussen. *LISA Metrology Systen - Final Report.* DTU Space / AEI Hannover / Axcon ApS, 2014

◆ A. Sesana, W.J. Weber, C.J. Killow, M. Perreur-Lloyd, D.I. Robertson, H. Ward, E.D. Fitzsimons, J. Bryant, A.M. Cruise, G. Dixon, D. Hoyland, D. Smith, J. Bogenstahl, P.W. McNamara, R. Gerndt, R. Flatscher, G. Hechenblaikner, M. Hewitson, O. Gerberding, S. Barke, N. Brause, I. Bykov, K. Danzmann, A. Enggaard, A. Gianolio, T. Vendt Hansen, G. Heinzel, A. Hornstrup, O. Jennrich, J. Kullmann, S. Møller-Pedersen, T. Rasmussen, J. Reiche, Z. Sodnik, M. Suess, M. Armano, T. Sumner, P.L. Bender, T. Akutsu, and B.S. Sathyaprakash. Space-based detectors. *General Relativity and Gravitation*, 46(12), 2014

◆ M Hendry, C Bradaschia, H Audley, S Barke, DG Blair, N Christensen, K Danzmann, A Freise, O Gerberding, B Knispel, et al. Education and public outreach on gravitational-wave astronomy. *General Relativity and Gravitation*, 46(8):1–11, 2014

◆ Yan Wang, David Keitel, Stanislav Babak, Antoine Petiteau, Markus Otto, Simon Barke, Fumiko Kawazoe, Alexander Khalaidovski, Vitali Müller, Daniel Schütze, et al. Octahedron configuration for a displacement noise-cancelling gravitational wave detector in space. *Physical Review D*, 88(10):104021, 2013

◆ P Amaro Seoane, Sofiane Aoudia, H Audley, G Auger, S Babak, J Baker, E Barausse, S Barke, M Bassan, V Beckmann, et al. The Gravitational Universe. *arXiv preprint arXiv:1305.5720*, 2013

◆ O. Gerberding, S. Barke, I. Bykov, K. Danzmann, A. Enddaard, J. J. Esteban, A. Gianolio, T. V. Hansen, G. Heinzel, A. Hornstrup, O. Jennrich, J. Kullmann, S. M. Pedersen, T. Rasmussen, Z. Sodnik, and M. Suess. Breadboard model of the LISA phasemeter. In G. Auger, P. Binetruy, and E. Plagnol, editors, *ASP Conference Series; 9th LISA Symposium*, volume 467 of *ASP Conference Series*, pages 271 – 275, 2012

◆ Gerhard Heinzel, Juan José Esteban, Simon Barke, Markus Otto, Yan Wang, Antonio F Garcia, and Karsten Danzmann. Auxiliary functions of the LISA laser link: ranging, clock noise transfer and data communication. *Classical and Quantum Gravity*, 28(9):094008, 2011

◆ Juan José Esteban, Antonio F. García, Simon Barke, Antonio M. Peinado, Felipe Guzmán Cervantes, Iouri Bykov, Gerhard Heinzel, and Karsten Danzmann. Experimental demonstration of weak-light laser ranging and data communication for LISA. *Optics Express*, 19(17):15937–15946, 2011

◆ S Barke, M Tröbs, B Sheard, G Heinzel, and K Danzmann. EOM sideband phase characteristics for the spaceborne gravitational wave detector LISA. *Applied Physics B*, 98(1):33–39, 2010

◆ M Tröbs, S Barke, J Möbius, M Engelbrecht, Th Theeg, D Kracht, B Sheard, G Heinzel, and K Danzmann. Fiber modulators and fiber amplifiers for LISA. In *Journal of Physics: Conference Series*, volume 228, page 012042. IOP Publishing, 2010

◆ M Tröbs, S Barke, Th Theeg, D Kracht, G Heinzel, and K Danzmann. Differential phase-noise properties of a ytterbium-doped fiber amplifier for the Laser Interferometer Space Antenna. *Optics letters*, 35(3):435–437, 2010

◆ M Tröbs, S Barke, J Möbius, M Engelbrecht, D Kracht, L d'Arcio, G Heinzel, and K Danzmann. Lasers for lisa: Overview and phase characteristics. In *Journal of Physics: Conference Series*, volume 154, page 012016. IOP Publishing, 2009

◆ Simon Barke, Michael Trös, Benjamin Sheard, Gerhard Heinzel, and Karsten Danzmann. Phase noise contribution of EOMs and HF cables. *Journal of Physics: Conference Series*, 154(1):012006, 2009

◆ M Tröbs, S Barke, J Möbius, M Engelbrecht, D Kracht, L d'Arcio, G Heinzel, and K Danzmann. Lasers for LISA: Overview and phase characteristics. In *Journal of Physics: Conference Series*, volume 154, page 012016. IOP Publishing, 2009

◆ Simon Barke. Inter-Spacecraft Clock Transfer Phase Stability for LISA. Diploma thesis, Leibniz Universität Hannover, Institute for Gravitational Physics, 2008

◆ Sebastian Steinlechner, Simon Barke, Jessica Dück, Lars Hoppe, Raoul Amadeus Lorbeer, Markus Otto, Aiko Samblowski, and Tobias Westphal. Der lifter-ein flugobjekt mit ionenantrieb. *PhyDid A-Physik und Didaktik in Schule und Hochschule*, 2(7):20–26, 2008

PROJECT DOCUMENTS

Availability can be inquired through the European Space Research and Technology Centre (ESTEC) of the European Space Agency (ESA) under ITT Reference Number AO/1-6238/10/NL/HB.

- ◆ LISA Metrology System Team. Metrology system BB requirements specification. Technical Report TN1-1:AO/1-6238/10/NL/HB, European Space Agency, 2011

- ◆ LISA Metrology System Team. Metrology system consolidated architecture. Technical Report TN1-2:AO/1-6238/10/NL/HB, European Space Agency, 2011

- ◆ LISA Metrology System Team. Metrology system development model design description and test results. Technical Report TN1-3:AO/1-6238/10/NL/HB, European Space Agency, 2011

- ◆ LISA Metrology System Team. Metrology system test plan. Technical Report TN1-4:AO/1-6238/10/NL/HB, European Space Agency, 2011

- ◆ LISA Metrology System Team. Test set-up design report. Technical Report TN1-5:AO/1-6238/10/NL/HB, European Space Agency, 2011

- ◆ LISA Metrology System Team. Metrology system design report. Technical Report TN2-1:AO/1-6238/10/NL/HB, European Space Agency, 2012

- ◆ LISA Metrology System Team. Manufacturing Drawings. Technical Report TN2-2:AO/1-6238/10/NL/HB, European Space Agency, 2012

- ◆ LISA Metrology System Team. Metrology system as built design report. Technical Report TN3-1:AO/1-6238/10/NL/HB, European Space Agency, 2012

- ◆ LISA Metrology System Team. Test Set-up commissioning report. Technical Report TN3-2:AO/1-6238/10/NL/HB, European Space Agency, 2012

- ◆ LISA Metrology System Team. As Built Configuration List. Technical Report TN3-3:AO/1-6238/10/NL/HB, European Space Agency, 2012

- ◆ LISA Metrology System Team. Metrology system test report. Technical Report TN4-1:AO/1-6238/10/NL/HB, European Space Agency, 2013

- ◆ LISA Metrology System Team. Lessons learnt summary report. Technical Report TN5-1:AO/1-6238/10/NL/HB, European Space Agency, 2014

- ◆ LISA Metrology System Team. Metrology system and GSE User Manual. Technical Report TN5-2:AO/1-6238/10/NL/HB, European Space Agency, 2014

- ◆ LISA Metrology System Team. Road map for Metrology system Development up to Flight Model (and ROM quotation). Technical Report TN5-3:AO/1-6238/10/NL/HB, European Space Agency, 2014

CONFERENCE TALKS (SELECTION)

- Talks and presentations at the Laser Interferometer Space Antenna (LISA) Symposiums in Barcelona (2008), San Francisco (2010), Paris (2012), and Florida (2014)
- Various talks at the annual meetings of the German Physical Society (DPG) and the Committee on Space Research (COSPAR)
- Closed working group meetings, i. a. at Caltech Campus in Pasadena, CA, USA, and at ESA's European Space Research and Technology Centre (ESTEC) in Noordwijk, Netherlands.

OUTREACH & PUBLIC TALKS (SELECTION)

- Creation of scientific and educational videos, e.g. Gravity Ink. series and comprehensive Athena/eLISA Teaser
- Coordination of exhibitions and related public talks at trade shows (i. a. IdeenExpo 2011) and conferences (i. a. DPG-AMOP 2013)
- Contestant of multiple national and international Science Slam (short presentation) competitions between 2011 and 2014
- Ceremonial address for award presentation at the Scion DTU Science and Technology Park, Copenhagen, Denmark, in 2011

THESIS SUPERVISION

- Sebastian Hild, Universität Bonn / University of Florida: LASER Frequency Stabilization Using Heterodyne Interferometry for the LISA Spacecraft Mission (Bachelor's thesis, 2009)
- Amrit Pal Singh, Leibniz Universität Hannover / Albert Einstein Institute: Computation of Phase-Stability Over Temperature - Thermal Testing of Microwave Cables for the Laser Interferometer Space Antenna (Bachelor's thesis, 2010)
- Daniel Edler, Leibniz Universität Hannover / Albert Einstein Institute: Measurement and Modelling of USO Clock Noise in Space Based Applications (Bachelor's thesis, 2014)

COLOPHON

This document was typeset in LaTeX using a modified version of the typographical look-and-feel *classicthesis* developed by André Miede (GNU GPL).

Two-dimensional plots were created using *gnuplot*, a command-line program, and converted to the PDF file format by *librsvg*, a cross-platform graphics library (GNU GPL).

Three-dimensional plots were created using R, a programming language and software environment for statistical computing and graphics (GNU GPL).

Spectral densities were computed by *LPSD* based on 'Improved spectrum estimation from digitized time series on a logarithmic frequency axis' by Michael Tröbs and Gerhard Heinzel [154, 155].

Models were rendered using *Autodesk* "Inventor Professional 2012". Graphics were edited using *Adobe* "Illustrator CC 2014" and "Photoshop CC 2014". Videos were compiled using *Adobe* "Premiere Pro CC 2014" and "After Effects CC 2014".

Bibliographic information published by the Deutsche Nationalbibliothek
The Deutsche Nationalbibliothek lists this publication in the Deutsche Nationalbibliografie; detailed bibliographic data are available on the Internet at http://dnb.dnb.de.

© 2015 Simon Barke
SBP – Simon Barke Publishing
Production and distribution: Ingram Content Group Inc.

ISBN: 978-3-946068-08-2

www.ingramcontent.com/pod-product-compliance
Lightning Source LLC
Chambersburg PA
CBHW051838210326
41597CB00033B/5703